Water Infrastructure Protection and Homeland Security

FRANK R. SPELLMAN

GOVERNMENT INSTITUTES
An imprint of
The Scarecrow Press, Inc.
Lanham, Maryland • Toronto • Plymouth, UK
2007

Published in the United States of America
by Government Institutes, an imprint of The Scarecrow Press, Inc.
A wholly owned subsidiary of
The Rowman & Littlefield Publishing Group, Inc.
4501 Forbes Boulevard, Suite 200
Lanham, Maryland 20706
http://www.govinstpress.com/

Estover Road
Plymouth PL6 7PY
United Kingdom

British Library Cataloguing in Publication Information Available

Library of Congress Cataloging-in-Publication Data

Spellman, Frank R.
Water infrastructure protection and homeland security / Frank R. Spellman.
p. cm.
ISBN-13: 978-0-86587-418-3 (pbk. : alk. paper)
ISBN-10: 0-86587-418-2 (pbk. : alk. paper)
1. Waterworks—Security measures—United States. 2. Terrorism—United States—Prevention. I. Title.
TD485.S665 2006
363.325'936361—dc22 2006028862

∞™ The paper used in this publication meets the minimum requirements of American National Standard for Information Sciences—Permanence of Paper for Printed Library Materials, ANSI/NISO Z39.48-1992.
Manufactured in the United States of America.

For Revonna,
the ultimate geoduck

Contents

Preface

In January 1992, while serving as the environmental health and safety manager of a large sanitation district (wastewater treatment), I took it upon myself to develop a written protocol for compliance with OSHA's Process Safety Management (PSM) Standard, 29 CFR 1910.119. Initially, during the protocol development stage, I did not receive strong upper management support and received even less encouragement. It took me five years to develop and implement PSM (the implementation date, required by OSHA/U.S. EPA was May 1997). However, I fully understood the need and the urgency to comply with the new OSHA standard—I became a strong supporter of PSM. This transition from the dread of trying to comply with "just another one of those regulatory standards" to understanding the need to comply did not occur overnight. I bought into the process slowly. During the development of the PSM protocol, the EPA jumped on OSHA's bandwagon. The EPA saw the efficacy of PSM, with a few modifications. Thus, the EPA "borrowed" OSHA's PSM standard, modified it, and called it Risk Management Program (RMP), 40 CFR Part 68 regulations. The PSM and RMP regulations were developed to help organizations (companies, plants, manufacturers, etc.) further protect their employees (OSHA), workplace (OSHA), public (EPA), surrounding community (EPA), and the environment from releases of hazardous materials. I liked PSM/RMP so much and bought into its requirements so thoroughly that I said so in the book *A Guide to Compliance for PSM/RMP*, published by Technomic in 1998.

The obvious question is, what does PSM/RMP have to do with this book, *Water Infrastructure Protection and Homeland Security*? The simple answer is that pre-9/11, they actually had quite a lot to do with each other. The compound answer is that post-9/11—well, the compound answer is complicated but will become clear as you proceed through the text.

Moreover, in using this text, it will also become quite clear that the events of 9/11 have forced an increased focus on water/wastewater infrastructure in the United States.

This book was written as a result of 9/11 and in response to the critical needs of water/wastewater plant managers, plant engineers, design engineers, utility managers, and anyone with a general interest in the security of their water/wastewater facility as a resource on the security of water/wastewater systems. It is important to point out that water/wastewater systems cannot be made immune to all possible intrusions/attacks; thus, it takes a concerted, well-thought out effort to incorporate security upgrades in the retrofitting of existing systems and careful security planning for all new construction sites. These upgrades or design features need to address issues of monitoring, response, critical infrastructure redundancy, and recovery to minimize risk to the facility/infrastructure.

Water Infrastructure Protection and Homeland Security presents common sense methodologies in a straightforward manner. Why so blunt? At this particular time, when dealing with security of workers, family members, citizens, and society in general—and with our very way of life—politically correct presentations on security might be the norm, might be expected, and might be demanded. Frankly, my view is that there is nothing normal about killing thousands of innocent people; mass murders certainly should not be expected, and the right to live in a free and safe environment is a reasonable demand.

This text is accessible to those who have no experience with security. If you work through the text systematically, an understanding of and skill in security principles can be acquired—adding a critical component not only to your professional knowledge but also giving you the tools needed to combat terrorism in the homeland—our homeland.

Prologue

Call me Jake. Some time ago—never mind how long precisely; my time is limited—having formed a group of like-minded American patriots, I decided to put my master plan into effect. The plan was simple enough—brilliant actually. In several cities throughout the United States, my team members and I were to plant dirty bombs in underground sewage tunnels and vaults. These underground sewage conveyors, all of them large enough to allow an army of my fellow compatriots easy access and room to operate, were carefully, painstakingly selected. Each was located in a major U.S. city, directly under federal courthouse complexes. Many of these complexes also included high-rise condos, banks, restaurants, bookstores, and so on. A high kill rate was paramount, critical, necessary, and part of our plan, of course.

I had worked for a large sanitation district in an average-sized city for about 14 years. During that period, I noticed how slack things were in the organization. For example, workers (in many instances) showed up for work when they wanted to, and when finally at their workstations, sat at their computers most of the day playing solitaire, especially those in the water quality division. The general manager was promoted not based on ability to lead or manage but solely on 25 years of longevity and backstabbing. In regard to his backstabbing, I can only say that the methodology he employed to assure his ascension would not have made Niccolò Machiavelli roll over in his grave; instead, it would have made him sit up with a bony smile, clap bones together, and cheer breathlessly. I called this general manager the Space Cadet in charge or simply Space Cadet Inc. Space Cadet Inc. was a man of little ability but great ambition, which brings to mind that Oscar Wilde quote: "Ambition is the last refuge of the failure."

The fact is, I worked at the sanitation district under the guidance of a total misfit. For example, right after the 9/11 fiasco, Space Cadet Inc. panicked. On the day after, September 12th, he was waiting for me outside my office at 5:00 a.m. As I opened the office door (I was the organizational occupational safety and health manager), Space

Cadet Inc., in a frightened, hyped-up, quavering voice, told me he had an important assignment for me. "Jake, after what happened yesterday, we need to ensure that we are covered in this organization. I want you to form a team (don't all dysfunctional managers form a team during a crisis? Part of their modus operandi) and figure out how to prevent a 9/11 from occurring here."

While sitting behind my desk and watching Space Cadet Inc. attempt and finally succeed in fitting his overstuffed presence into the only other chair in my dinky office, I thought to myself, "Why in God's name would terrorists fly airplanes into sewage treatment plants? Sure, the terrorists are crazy, but are they also stupid?"

Space Cadet Inc. said, "When you form the team, Jake, I want you to determine how to secure the district and how to ensure we're covered. I can see those press headlines now: 'Local Sanitation District Wastewater Treatment Plants Struck by Terrorists!' We can't have that. You know how those press bums are; they will have a feeding frenzy. They will eat me, ah, us alive. You have a reputation for getting it done, Jake, so get it done!"

You might have thought I was somewhat incredulous with/about Space Cadet's statement, actions, and weird behavior. Actually, I was not. It is true that the more he spoke, the deeper that familiar sensation of cold, sharp steel penetrating my back felt. But, I was used to it. I mean, think about it: When you have a space cadet with a normal operating blood pressure of 200 over 150, what else can you expect?

Anyway, I told Space Cadet Inc. that I would move on his directions.

In the following weeks, I formed a team (members of which chose not to show up at any of the meetings because the whole mess to them was just that—a mess). Thus, we had no meetings. Instead, the meetings consisted of one person, me. I did, on my own, think through what was needed, read the U.S. EPA's early guidelines on homeland security, and make many decisions.

For example, I decided on certain items that were needed to secure the nine major wastewater treatment plants, maintenance centers, pumping stations, and main office complexes. The security steps I incorporated (many at the same time) are listed below:

- **Awareness:** The key to any new program is to make sure the employees are aware of the problem and the new changes. Initially this was easy; all the hype over 9/11 was still fresh in their human microprocessors.
- **Vulnerability Assessment:** I conducted a thorough walk-through of all district worksites, including treatment plants, pumping stations, maintenance facilities, and administrative offices, looking for vulnerable access points.
- **Needs List:** Based on the vulnerability study, I composed a list of needs. For example, I determined where video cameras, self-locking gates, warning signs, upgraded lighting, and magnetic card reader systems needed to be installed.

- **Presentation of Findings and Recommendations:** Space Cadet Inc. listened to all my findings and recommendations, but would not comment, one way or the other. Instead, he instructed me to put together a PowerPoint presentation and present it to the commission (the district's governing body), which provided oversight of district operations.
- **Commission Presentation:** I made a 20-minute presentation before the commission. All nine members sat there mesmerized and at the end voted unanimously to approve my recommendations. The chairperson then soundly chewed out Space Cadet Inc. and instructed him that whenever other district employees came before the commission, they could and should learn the proper presentation technique from me. I could feel Space Cadet Inc.'s hateful eyes staring at me in disbelief and also, again, that feel of cold, sharp steel penetrating deeper into my back.
- **Implementation:** It took me about four months to manage and oversee the (1) installation of a new magnetic strip-reading door entry systems at all major sites; (2) upgrading of fences, including the installation of the best razor wire ratepayer's money could buy; (3) installation of several video cameras at strategic locations throughout the district; (4) changing of several doors from standard types to security types; (5) installation of security windows in many doors to allow inside employees to see outside the door and onto the surrounding property before they opened the door; (6) removal of all brush, shrubbery, and trees in close proximity to buildings, which removed barriers that terrorists or others could hide behind; (7) hiring of a landscaping outfit to trim several overhanging tree limbs that would have allowed anyone outside the plant easy entry into our plant sites; (8) incorporation of cyber/SCADA computer upgrades; (9) installation of numerous 400-watt lighting fixtures along perimeter fences and chemical storage buildings, especially near gaseous chlorine and sulfur dioxide storage areas; (10) incorporation of a weekly audit program whereby I visited various plant sites and pumping stations at night to ensure that all gates were closed and secure and that fences were intact; and finally (11) updating of the district's Emergency Response Plan for all contingencies—including the contingency that a fully loaded airplane could crash into a sewage basin.
- **Outside Contractors/Visitors:** Outside contractors and visitors have the tendency to wander when and where they want to if not escorted or indoctrinated on the off-limits or restricted areas. I also have found that contractors, working a major retrofit or new construction at our plants, need access to and from the plant 24/7. To control ingress/egress and traffic flow of contractors, I designated specific access points and manned these locations with radio-carrying professional security guards.

I was able to accomplish the above for one main reason: I had unlimited access to ratepayer funds. Initially I told Space Cadet Inc. that I needed $50,000, but in the end

I spent more than $250,000. I was able to spend all this money with no questions asked. Why? Along with the dysfunctionalism of Space Cadet Inc., the lower-level managers, whose responsibilities included overseeing operations and ensuring accountability, were also dysfunctional—very lax enforcers of de facto policies. Thus, I went on a spending spree and enjoyed spending money that did not belong to me. I bet I could have used ratepayer money to purchase a new sports car, a sailboat, or even a fully equipped bowling alley and no one would have cared! But of course, I spent the money for security upgrades only. I am an honest terrorist, and besides, I don't bowl.

Anyway, fast-forward to the near present. After all those years of exposure to dysfunctional management, working for and with space cadets, I grew weary of my security responsibilities, and of any and all responsibilities!

So, from my diatribe to this point, I think you can understand my motivation in striking against various sanitation districts throughout the lower 48 states. But actually there was/is more to it than just that. It all had more to do with targets of opportunity—easy targets, so to speak.

My compatriots looked at water utilities first. We certainly had a huge stockpile of chemicals and biologicals that we could have used to contaminate (poison) water supplies throughout the country. In America, you can buy literally anything; all one needs is the money. Chemicals and biologicals were not the problem. No, the problem was bang for the buck. Sure, we could have easily poured our death and destruction into local reservoirs and other water supply sources. We could have killed hundreds and terrified thousands. But that is not enough. As a case in point, consider the destruction of the federal building in Oklahoma City. A few hundred folks killed and a building destroyed—that amounted to nothing. Then there was 9/11, which killed almost 3,000 and caused major damage to infrastructure—who cares? Nothing more than a black eye to the American psyche. No, sir, I wanted more—I wanted to totally disable America, to bring her to her knees. That's what I wanted, but unfortunately, we got caught before we could send the ultimate message. In the end, you can't win them all, but we only need to win one.

I had 33 dirty bombs and crews ready to enter underground sewers in major U.S cities. It was the perfect plan, orchestrated by the ultimate genius, thank you very much. But we got caught in the act by a bunch of dysfunctionals.

You have to understand that I fully understood the Achilles' heel of wastewater treatment systems: the sewers. This fact came to me when I was doing my post-9/11 security upgrades in the district. I took this information to Space Cadet Inc., but he was unimpressed and said, "The sewers belong to the cities. That's their problem. If anything goes wrong, it is them that will swing, not me, ah, I mean us. All we do here is collect their sewage and pump it to our treatment plants."

It's hard to argue with Space Cadet Inc.'s logic. The fact is, all we did was intercept the city's flow and pump it to our plants. Anyway, this fact gave me the idea on how to attack downtown areas, especially federal courthouses, using underground sewers—out of sight, out of mind.

My plan was perfect; there was no way it could not work. Big bang for the buck!

However, in putting my plan into effect, I began to lift that first manhole cover, and before I could wrestle it off, they were on me like ticks on a dog. The commie feds, a glorified SWAT team, and state police were there, too. After being roughed up a bit and forced to rat out my compatriots (personal survival is much more important to me than my compatriots), the feds were able to stop the other 32 teams before they even placed one dirty bomb where it was needed.

I hated to rat out my compatriots; rats are the lowest form of scum. However, I am not crazy; I am not a suicide bomber. My theory is that radical suicide bombers do themselves in for one reason and one reason only. Not for religious beliefs. They want to die because they can't face failure if their attacks fail. Me? I know failure is possible. Therefore, suicide is out of the question. I use failure as a lesson learned. Besides, if I had been successful, I wanted to be around to glory in the success.

Anyway, subsequently, I have suffered through two years in court, having to put up with the same system I hate so much. I only wish I had known what I eventually found out in court. Mainly, that the manhole cover I tried to lift was alarmed—some kind of proximity switch or microchip device all tied together with some type of fiber optic, or whatever. The point is that when I lifted that cover all kinds of bells and whistles went off at what they called their central control station, which also was tied directly to the police station one block from the manhole. Sometimes things just don't work out, even for us geniuses. It turned out that I had selected the wrong sanitation district to attack; the other 32, according to court testimony, were not alarmed. Just my luck!

Now I sit here waiting to be transported to that supermax prison in Colorado. Yeah, I have heard all about that place. It is designed to house violent prisoners or prisoners who might threaten the security of the guards or other prisoners. Supermax is the place where they lock down prisoners 23 hours a day. There is a small triangular recreation area, known as "the dog run," where solitary prisoners can occasionally get a glimpse of sky. Food is delivered through a slit in the cell door. All the cell furnishings are made of cement. Can you believe that? Prisoners don't leave their cells to see a lawyer, a doctor, or a prison official; those visitors come to the cell. Motion detectors and cameras monitor every move. The prison walls and razor-wired grounds are patrolled by laser beams and guard dogs. Prisoners are not rehabilitated at Supermax. No sir, they vegetate.

Stark Supermax conditions don't bother me, however. No, what really torques my jaws is the fact that they are going to place a red-blooded American like me, who did not kill anyone, in the same prison with scum bags like that idiot Moussaoui; Ramzi Yousef, the mastermind of the 1993 World Trade Center blast; "Unabomber" Theodore Kaczynski; Terry Nichols, an accomplice in the 1995 Oklahoma City bombings; Richard Reid, the so-called shoe bomber; and Eric Rudolph, who bombed abortion clinics and the Atlanta Olympics.

Well, I guess I will learn to live with my plight. Do I have any other choice? If I had another choice, I would be outside this holding cell enjoying a hot cup of Starbuck's best and some of those tasty geoducks (pronounced gooey ducks, thank you very much).

But, alas, I hear them coming for me. I hope this note will get to the public out there in la la land; I am going to hand it to my lawyer.

At the moment, only one thing haunts me: How did those sewer folks get smart enough to alarm their manhole covers? I know of no book on the subject. If I were free, I would write a book about water/wastewater security. No one could write it better. I guess someone else will have to write that book.

I just don't get it. My fate is in the hands of the idiots I detest the most. Why me? I just don't get it!

Come to think of it, a better question might be: Do you get it?

Jake
Holding Cell #3
Seattle Federal Courthouse

1

Introduction

You may say Homeland Security is a Y2K problem that doesn't end January 1 of any given year.

—*Governor Tom Ridge*

Governor Tom Ridge, a U.S. political figure who served as a member of the U.S. House of Representatives (1983–1995), governor of Pennsylvania (1995–2001), assistant to the president for Homeland Security (2001–2003), and the first U.S. secretary of homeland security (2003–2005), had it right; Homeland security is an ongoing problem that must be dealt with 24/7. Simply, there is no magic on-off switch that we can use to turn off the threat of terrorism in the United States or elsewhere.

The threat to our security is not only ongoing but also universal, including potential and real threats from within—from our own citizens. Consider the American Timothy McVeigh, for example, who blew up the government building in Oklahoma City in 1995, killing almost 200 people, including several children. McVeigh, who bombed the building in revenge for the FBI's Waco, Texas, raid, thought the army (he was a former decorated U.S. Army veteran) had implanted a chip in his body to track his movements, according to reports.

It is interesting to note that McVeigh, who was no doubt suffering from some type of severe disturbance, acted primarily alone. Actually, McVeigh is the exception that proves the rule—most terrorist acts on America are planned by a group beforehand, but, this is not always the case. For example, consider the scenario that follows. Note that this incident occurred in 1991 pre-McVeigh, pre-first attack on the World Trade Center, and pre-9/11.

Revenge Is Mine Sayeth Daniel

Daniel, tall, lean, mean, and evil, stood there on the rail tracks outside the wastewater treatment plant's western fence line, peering through the 8-foot chain-link fence surrounding the plant site at the 55-ton rail tank car inside the plant. When not checking on the tank car, his gaze flashed from right to left, left to right, and back to the railcar. He was being careful, watchful, and deliberate. He was well aware of the plant operator's schedule, her hourly rounds; he had worked the same plant, same sampling route, and same job for five years before his "unlawful" termination.

He thought, "Those frog-faced managers said I was greedy, heartless, predatory, unethical, lazy, buffoonish, and incompetent— a real basket-case, one of them said. Well, I will show them."

As the last rays of sunshine struck his back (he did not feel those warming rays; his body was cold, dry, and ready; warming had no effect), he glanced down to the cardboard box at his feet. Then, having heard something, he crouched a bit and glanced inside the plant again—nothing stirred, not even his breath, just lengthening shadows. Upright again, he glanced at the box, filled with the 12 bottles of yellow-gold liquid. The sight of the bottles warmed his coldness a bit. As he stood there waiting for the right moment, for the sun to meet the horizon, his long, sharp-looking face almost grimaced into a smile as he thought about those 12 bottles of death at his feet.

Just a week earlier, when he was convinced his master plan would work (failure was not an option), he had fashioned his crude incendiary weapons. The Molotov cocktails (Daniel liked to call them homemade frag) were carefully put together using tall, glass bottles partly filled with gasoline with a touch of sugar to help the gasoline cling to the target. The mouth of each bottle was stopped up with a cork, and a cloth rag was fixed securely around each mouth.

The target of these cocktails? The 55-ton railcar, fully loaded, of course, with deadly chlorine gas. In Daniel's mind, this was the perfect scenario. He could already foresee that yellow-green mushroom cloud of death crossing the plant site, aided by the wind, which seemed to be freshening as the sun touched the horizon.

Looking all around to ensure that he was alone, no one was near the railcar, no one was behind him, and his path of escape was clear, he pulled the whiskey flask filled with gasoline and the disposable lighter from his coat pocket. Quickly, he poured gas on each of the 12 bottle wicks and put the closed flask back into his pocket. He picked up the first bottle, lit the rag, looked around, and threw the lighted bottle over the fence toward the railroad car. He missed. The first bottle landed in the tall grass, next to a stack of wood and junk; it just sat there, though he could see the lighted wick in the darkening plant. He lit another wick and threw. He repeated this procedure several times until the box was empty. He took one last glimpse at his work and noticed that two or three of the bottles had actually struck the metal of the tank car, bursting and flaming instantly. His work was done. It was time to retreat quickly; he did.

Meanwhile, during the same time period . . .

The last rays of the setting sun touched the running waters of the large river that coursed its way through the downtown region of a large metropolitan area. A plant operator at the 100-million-gallon-per-day wastewater treatment plant, located in the city along the same river bank, was walking the plant site making her rounds. Stopping at a sample point, she pulled a sample, then deposited the sample and its bottle into the carrying tray. Grabbing the tray, she proceeded down the long, winding stairway from the #9 secondary clarifier. At the foot of the stairs she stepped onto the gravel path and proceeded toward the final effluent sampling point.

She walked along the path and then turned to her right. Straight ahead, just behind the 55-ton chlorine rail tank car, she noticed an orange glow.

Recognizing the orange glow for the brush fire that it was, the operator dropped the sample tray and ran the 250 feet to the plant's main control center to alert the shift lead operator about the brush fire near the chlorine tank car. While she dialed 9-1-1, the lead operator activated the site emergency alarm and then used his portable radio to direct a plant assistant operator to meet him at the tank car.

With the plant's emergency alarm siren wailing in falsetto throughout the plant site and the neighborhood, the lead operator and assistant were standing together approximately 100 feet from the chlorine tank car. They could see the brush fire was growing in strength; it was being fed by a brisk wind.

The lead operator wondered what to do next—really, what to do *first*—but his mind was blank. Then it cleared, and a series of instantaneous thoughts entered his mind. First, he understood the gravity of the situation: a growing fire was about to engulf 55 tons of chlorine. Second, he realized there was no way he and the assistant operator could move the tank car out of harm's way, though it appeared to him that the tank car was already engulfed in flames. Third, he realized the spur line the tank car was positioned on was heavily overgrown with brush (just a few weeks earlier he had directed a former employee, Daniel, to cut the overgrowth and remove the pile of wood near the railcar). As the lead operator and assistant advanced a few yards closer to the tank car, the lead operator noticed another problem. A plant maintenance crew had stacked a pile of wooden cement forms next to the spur line, within a few feet of the tank car; these forms were on the same side of the car where fire was quickly approaching from the end not already aflame. The lead operator knew he would have to act fast to prevent an extremely dangerous situation, the total engulfment of the tank car.

He had to do something.

He did.

The lead operator directed the assistant to go over to the nearest building, the non-potable water pump house, and bring a fully charged fire hose back to the fire with him. Then the lead operator darted off in the opposite direction, toward the chemical handling building, to get another fire hose.

About five minutes later, both the operators, manning two fully charged 1.5-inch fire hoses, approached the tank car from the side opposite where the brush fire had already reached the spur and had ignited the wood cement forms; the fire was beginning to grow, to burn with purpose.

Standing to one side with his fire hose directed at the midsection of the chorine tank car body, while at the same time, the assistant, at the other end of the tank car, began directing a steady stream of nonpotable water (NPS) in the same general area, the operator could feel the heat from the growing fire.

A few minutes later, the fire was burning the entire stack of wood cement forms, generating such intense heat that both operators had to move back a few feet from the car. At that same moment, the tank car emergency relief valve, having reached its design activation pressure, lifted, releasing a controlled stream of deadly chlorine gas to the atmosphere.

Along with the plant alarm siren, the operators could hear the sirens of emergency response vehicles approaching the plant site as the tank car emergency relief valve failed completely. A bad situation got worse. Now, instead of releasing a controlled amount of chlorine gas, the entire contents of the tank car were escaping full force, a steady stream of yellow-green death, all 55 tons.

About six-tenths of a mile away from the treatment plant, the fire, and the escaping gas, at about the time the emergency relief valve on the 55-ton chlorine tank car had failed, several evening classes at the university were in progress. The 400 college students had heard the plant site emergency alarm siren and then the other sirens as emergency vehicles raced by the university campus toward the wastewater treatment plant, but they paid them no mind. They had heard these alarms several times before; emergencies in the central city were a common occurrence.

Meanwhile, back at the plant, the operators were now fully engulfed by the chlorine's yellow-green cloud of death; they were about to take their last breaths. In the plant control room, the operator who had discovered the fire and sounded the alarm had been busy. Not only had she notified the authorities about the emergency at the plant, but she had also called the plant manager and chief operator and filled them in about the fire.

When she heard the emergency responder's sirens, she ran outside just in time to see the fire department and local hazardous materials (hazmat) team enter the plant site through the front gates. However, because darkness had set in, she did not see the dense yellow-green cloud of chorine gas and she walked right into it. Instantly overcome by the choking chlorine gas, she fell to her knees, coughing and gasping for air. Instead of air, she filled her lungs with deadly chlorine gas; she died five minutes later.

The emergency responders were not familiar with the plant site they had entered; they had not been invited to tour the plant site to learn the layout of the site. However, from information provided to the 911 operator, the emergency responders did know about the fire and the 55-ton tank car of chlorine. From knowledge they had gained through their training, they understood the danger involved with chlorine gas and fire. What they did not know, however, was that the fire had already reached the chlorine tank car and that the tank car emergency relief valve had already activated and released its entire load into the atmosphere.

However, these firefighters and hazmat responders knew their jobs. They had been properly trained in hazardous materials emergency response procedures; therefore, as they entered the plant site they used caution. They were also alert enough to recognize, with the help of their spotlights, that the yellowish-green cloud of death was moving directly toward them.

It didn't take long before the Fire Captain in charge of this emergency situation gave the order for his responders to retreat to safer ground; they did.

Meanwhile, at the university, two blocks from the treatment plant—less than 600 feet away—class had ended and students poured out of their classrooms. Several students left the building to go home. Others stepped outside for a few minutes to smoke cigarettes before the start of their next class.

About the time the emergency responders were exiting the plant site to set up a command post in a safe zone (approximately 1.6 miles from the plant), the same brisk wind that had steered the fire toward the tank car and stoked the fire there was pushing the ground-hovering poisonous gas toward the front entrance of the college buildings where students were joking and smoking.

Within a few minutes, the chlorine cloud came face-to-face with the college students.

Several of the students survived the chlorine gas with only minor respiratory irritation. A few were more seriously affected; they were rushed to a local hospital. An even smaller number of students, those in the parking lot fronting the school, were more profoundly affected; later, three of these students died.

During daylight the next day, after the chlorine had dissipated and no longer was an issue, and for several days after this incident, investigators and other interested viewers had little trouble following the path the deadly chlorine cloud had taken. Its 1.5-mile path was clearly marked by dead grass, flowers, insects, bushes, trees, cats, dogs, raccoons, squirrels, and ducks.

After an extensive criminal investigation, Daniel was charged, tried, and executed in the electric chair. This entire process was complete by 2001(Spellman, 1997).

Daniel's Revenge: Aftermath

The incident just related points to the purpose of this text; namely, to emphasize the importance of water infrastructure systems to be properly prepared for any contingency. In 1990–1991 (along with other American industries involved in hazardous materials production, shipping, storing, and usage), this was the intent of the Occupational Safety and Health Administration's (OSHA) Process Safety Management Standard (PSM) 29 Code of Federal Regulations (CFR) 1910.119, which was, as mentioned earlier in the prologue, promulgated in 1992 and to be fully complied with by May of 1997.

OSHA's primary purpose (a lofty but important goal) is to protect the worker from accidents and illnesses in the workplace. PSM is simply an extension of this effort. It should be pointed out, however, that OSHA standards, like PSM, are designed, almost exclusively, to protect workers within workplace fence lines. Herein lies the problem with PSM. Hazardous materials spills, accidental or intentional, often cross workplace fence lines into surrounding neighborhoods, affecting all three environmental media: air, water, and soil. Thus, while the Environmental Protection Agency (EPA) saw the benefit of PSM and applauded OSHA's efforts in this critical safety area, it also saw the drawback. Simply, hazardous materials incidents that begin in the workplace are likely to spread their lethal effects beyond the fence line. This is the primary reason that the EPA borrowed many of the tenets of PSM and used that material to develop its own regulation, Risk Management Planning (RMP), 40 CFR Part 68 regulations. Both of these regulations are designed to prevent and/or detail the proper mitigation procedures to be utilized during hazardous material incidents. The key point to remember is that PSM/RMP not only combined functions to set requirements for onsite emergency response planning and training, they also pointed out the need for emergency response planning for offsite consequences.

What does all this information about PSM/RMP have to do with protecting water/wastewater infrastructure? Good question. For those already familiar with the tenets of these important regulations, the answer is obvious. That is, in combination, these regulations set the basis and became the vital foundation for subsequent directives related to post-9/11 homeland security and protection of vital infrastructure, including water/wastewater infrastructure, from terrorism—domestically generated or otherwise.

To gain a better sense of the positive impact that compliance with PSM and RMP can have on facilities that use or produce covered hazardous materials in their processes (including water/wastewater systems), consider the previous disaster scenario from a different point of view. That is, the preceding tank car incident, arguably a clear act of terrorism, could have developed in a much different manner if the elements and procedures required by both PSM and RMP had been in place and had been followed. This is not to say that in 1991 the event could have been prevented. I doubt that it could have been prevented because of our pre-9/11 mentality of that time frame. On the other hand, if PSM/RMP and present homeland security guidelines in effect today had been in effect in 1991, the event might have been prevented—nipped in the bud, guarded against.

In the first place, PSM and RMP require that all responsible parties survey their industrial complexes where covered chemical processes are employed and closely scrutinize these processes to determine if any of the 130-plus "Highly Hazardous Chemicals" listed in OSHA's PSM and/or any of the 140-plus "Extremely Hazardous Substances" listed in the EPA's RMP are stored, handled, used, or produced onsite (e.g., offgases).

In the chlorine tank car scenario, compliance with PSM and RMP would have ensured that the wastewater treatment plant had been thoroughly surveyed for processes using or producing highly hazardous chemicals and extremely hazardous substances. This survey would have determined, of course, that the plant site used deadly chlorine, a covered chemical, in its process. Moreover, the survey would have noted the normal quantity of chorine stored on the plant site at any given time. This survey would also have made note of the fact that the quantities of chlorine stored on the plant site exceeded the PSM and RMP threshold quantities (TQ) of 1,500 and 2,500 pounds, respectively, for chlorine. (Chlorine stored on the plant site in quantities at or above the threshold quantities requires the plant to meet all the mandatory compliance conditions required by PSM and RMP.)

Along with a complete survey of covered chemicals under PSM and RMP, the chlorine tank car incident plant site would have been required to complete the following actions:

1. Plant management would have been required to perform a Process Hazard Analysis (PHA) for the entire chlorine process, including the 55-ton rail tank car. Obviously, 55 tons of chlorine stored in the railroad tank car would have placed the treatment plant under the requirements of both regulations. Additionally, during this PHA, all other plant processes using or producing covered chemicals would have been audited and a determination would have been made as to whether these processes used or produced other highly hazardous chemicals or extremely hazardous substances.
2. Because the plant used chlorine at levels greater than the TQ level listed in PSM and RMP regulations, plant management would have been required to comply with all regulatory requirements as related in the regulations.
3. One of the important elements of the PSM and RMP regulations that the plant would have been required to comply with is onsite and offsite consequence analysis or modeling to assess potential onsite/offsite exposures.

4. Another PSM and RMP element is the requirement for emergency response planning. This emergency response planning is required for both onsite and offsite (public) areas.
5. Plant operator training in hazmat response is also required by PSM and RMP. The plant operators who responded to the fire and then the chlorine leak would have been trained on how to properly handle this incident. This particular element is absolutely critical in any emergency response incident. To answer this question, take another look at the fire and chlorine incident again from a different view.

Let us assume that the well-trained operators assume that the fire might cause the temperature and thus the internal pressure of the chlorine tank car to be raised to a dangerous level (yes, it is obvious this would be the case, but let us assume for a moment for the sake of argument). With this increase in internal tank car pressure, the tank car emergency pressure relief valve would lift and release chlorine to the environment. Under PSM and RMP, the operators would have been properly trained to respond to this situation in the correct manner. For example, when they first responded to the fire, they knew that the other operator was notifying the proper authorities; thus, professional help would have been on its way in a matter of minutes. Moreover, they also knew that by using the cooling water from the fire hoses, they might be able to cool the tank car to a level where it could have been maintained to a safe temperature level. As a matter of fact, it was what they did not know that produced the massive cloud of deadly chlorine gas that killed them and the other victims and sent the fire department and hazmat team in fast retreat.

If the plant operators had been properly trained in hazmat emergency response procedures, in the case of 55-ton chlorine tank car emergencies, they would have known some basic facts about tank cars and chlorine gas that could have brought about a different result. Again, for example, the operators would have been trained on how to properly cool a tank car involved in fire. If properly trained, they would have known that the 55-ton tank car was designed to withstand the effects of most fires. That is, chlorine tank cars must be constructed as specified by the U.S. Department of Transportation (DOT). The DOT specifies that all chlorine tank cars are to be protected with four inches of insulation. Thus, when the operators were directing their fire hose at the mid-end sections of the tank car, they were wasting not only their time but also cooling water; plus, they were providing little protection for the tank car's most vulnerable device: its pressure safety valve (PSV). Since the tank car's most vulnerable spot is its top housing (dome area), where the car PSVs are located, the lead operator should have directed his water at the dome area and the assistant operator should have used a nozzle on his hose that produced a fine mist spray to cover the lead operator and protect him against the heat of the fire. The plant operators should have known this basic, standard fire-fighting procedure. However, as was clearly demonstrated by their actions, they did not. They did not know the proper procedures because they had not been trained.

Proper compliance would have helped mitigate this incident. Moreover, proper compliance with PSM and RMP would have helped in another way. For example, full compliance with these regulations would have ensured that the local fire department and hazmat team would have been familiar with the layout of the wastewater treatment plant and had full knowledge of the plant site's chemical hazards. Most importantly, prior to this incident, plant site operators and emergency responders would have been required to meet, discuss, and walk though contingencies such as the tank car fire. The purpose of having such a meeting and discussion is to determine the best method to use in combating all possible chemical emergencies.

It should also be pointed out that if this plant site had been in full compliance with PSM and RMP, the local university administrators (as part of the regulations' requirements for the right of the public to know the chemical hazards they might be exposed to in the event of a plant emergency) would have understood the meaning of the shrill alarm that sounded at the plant site when the chlorine emergency first occurred. The college administrators would have had the knowledge they needed to properly react to the emergency and correctly evacuate the college premises.

As graphic and terrible as the example of the chlorine tank car incident just described is, it should be pointed out that hazmat incidents do not always result only because of fire. Along with the constant threat posed by fire, there are also chemical disaster situations that can occur due to human error in system operation and/or a malfunction in system equipment.

There are other emergency situations that must also be considered and planned for, including floods, hurricanes, earthquakes, tornadoes, volcanic eruption, snow/ice storms, avalanches, explosions, truck accidents, train derailments, airplane crashes, building collapses, bomb threats, riots, and sabotage.

Post-9/11, we need to add terrorism to the preceding long list of emergency situations. In regard to the chlorine tank car incident, we need to realize that long gone are those days when railroad safety meant avoiding derailments and accidents. Today we must consider and dodge terrorist attacks from Molotov cocktails as well as from armor-piercing bullets—and any other weaponry terrorists can get their hands on.

It is also important to remember that chemical emergency situations can easily reach beyond the boundaries of any industrial plant. This is to be expected, especially in this age of population explosion with its characteristic suburban sprawl. It is not unusual to find, for example, a water/wastewater treatment or other industrial plant that once was isolated from city dwellers and then later became surrounded on all sides by neighbors. The point is that when a chlorine spill occurs in an isolated area, there may be no cause for general alarm; however, when a spill occurs in a plant site such as the one described in the chlorine tank car incident, it should be clear that the purposes of PSM, RMP, the Patriot Act, homeland security directives, and other safety/security factors are far reaching—and absolutely critical to the survival of a free society.

WHAT IS TERRORISM?

If we were to ask 100 different individuals to define terrorism, it might be surprising to many and not to some that we would likely receive 100 different definitions. As a case in point, consider the following: If we were to ask 100 different individuals to describe the actions of Daniel in the railroad tank car incident, how would they describe him and his actions? You might be surprised. In 2000 and 2002, pre- and post-9/11, 100 randomly selected Old Dominion University environmental health juniors and seniors ("Generation Why" students ranging in age from 20 to 46 years old) were asked to read about Daniel's chlorine tank car incident and reply to a survey questionnaire. The two questions and the students' responses to this unscientific survey are listed in table 1.1.

Table 1.1. Student Perceptions of Daniel's Actions

Question 1: In your opinion, Daniel was . . .

	Number of Responses	
	Pre-9/11 (2000)	*Post-9/11 (2002)*
Crazy	22	12
A disgruntled former employee	10	2
Insane	32	4
Misguided	1	0
A cold-blooded murderer	3	5
A misfit	1	0
Deranged	6	4
A lunatic	5	6
A bully	17	10
A terrorist	3	57
Not sure	0	0
Totals	100	100

Question 2: In your opinion, Daniel's actions are best described as . . .

	Number of Responses	
	Pre-9/11 (2000)	*Post-9/11 (2002)*
Madness	45	5
Frustration	9	2
Desperation	4	0
Dysfunctional thinking	2	0
Legitimate concern	1	0
Threatening	6	1
Terrorism	5	79
Workplace violence	20	0
Not sure	8	13
Totals	100	100

Note: The student response descriptors were provided to the students by the instructor.

From the Old Dominion University survey, it is clear that the students' perceptions of Daniel's actions in the chlorine tank car incident shifted dramatically from pre-9/11 to post-9/11. For example, before 9/11, when asked to select the best pre-9/11 descriptor to characterize Daniel, "crazy" and "insane" ranked high; however, after 9/11, the students' perceptions shifted away from crazy and insane to "terrorist." Likewise, "madness" and "workplace violence" ranked high in the students' pre-9/11 descriptions of Daniel's actions; however, his actions post-9/11 overwhelmingly were described as "terrorism."

It is interesting to note that even though the responders from 2000 reported prior to both 9/11 and the September/October 2001 anthrax attacks, this student group's responses were still provided after such events as the World Trade Center attack of 1993 and Timothy McVeigh's 1995 mass murder attack of the occupants of the government building in Oklahoma City, Oklahoma. This may explain why the year 2000 students were somewhat reluctant to describe Daniel's subsequent actions as terrorism and/or to label him as a terrorist.

After studying this apparent anomaly (in the author's view) for several years, it has become obvious that terrorism, like environmental pollution, is a personal judgment call. Consider, for example, two neighbors living next door to a foul-smelling wastewater treatment plant. One of the neighbors works full time at the treatment plant, while the other neighbor works elsewhere. Each morning when the neighbor who does not work at the treatment plant steps outside to go to work, she has learned to hold her nose against the horrendous odor emanating from the plant site. There is absolutely no doubt in her mind that she lives next door to a pollution source. On the other hand, each time the other neighbor, the full-time employee of the wastewater treatment plant, steps outside his home to go to the plant to work his shift, he smells the same odor his neighbor does. However, when the plant employee smells the odor, he does not smell pollution; instead, he detects the sweet smell of money in the bank and job security.

Terrorism by Any Other Name Is . . .

From the preceding discussion, we might want to sum up terrorism as being relative, a personal judgment. But is it really relative? Is it a personal judgment? What is terrorism?

Take your choice. Seemingly, there is an endless list of definitions. Let's review a few of these definitions.

Standard Dictionary Definition of Terrorism

After reviewing several dictionaries, a fairly standard definition of terrorism is the unlawful use or threatened use of force or violence by a person or an organized group against people or property with the intention of intimidating or coercing societies or governments, often for ideological or political reasons.

America's *National Strategy for Homeland Security* defines terrorism as "Any premeditated, unlawful act dangerous to human life or public welfare that is intended to intimidate or coerce civilian populations or governments" (NSHS, 2006).

The U.S. State Department defines terrorism as "Premeditated, politically motivated violence perpetrated against noncombatant targets by subnational groups or clandestine agents" (U.S. Congress, 2005).

The FBI defines terrorism as "The unlawful use of force or violence against persons or property to intimidate or coerce a Government, the civilian population, or any segment thereof, in furtherance of political or social objectives" (FBI, 2006).

Note that the FBI divides terrorism into two categories: domestic, involving groups operating in and targeting the United States without foreign direction; and international, involving groups that operate across international borders and/or have foreign connections.

At this point the obvious question is, do you now know what terrorism is? That is, can you definitely define it? If you can't define it, you are not alone—not even the U.S. government can definitively define it. Maybe we need to look at other sources.

The following is Osama bin Ladin's view on terrorism: "Wherever we look, we find the U.S. as the leader of terrorism and crime in the world. The U.S. does not consider it a terrorist act to throw atomic bombs at nations thousands of miles away [Japan during World War II], when those bombs would hit more than just military targets. Those bombs rather were thrown at entire nations, including women, children, and elderly people" (Bergen, 2002).

The following is another view (court testimony) on terrorism from Ramzi Ahmed Yousef, who helped organize the first terrorist attack on the World Trade Center: "You keep talking also about collective punishment and killing. Innocent people to force governments to change their policies; you call this terrorism when someone would kill innocent people or civilians in order to force the government to change its policies. Well, when you were the first one who invented this terrorism.

"You were the first one who killed innocent people, and you are the first one who introduced this type of terrorism to the history of mankind when you dropped an atomic bomb which killed tens of thousands of women and children in Japan and why you killed over a hundred thousand people, most of them civilians, in Tokyo with fire bombings.

"You killed them by burning them to death. And you killed civilians in Vietnam with chemicals as with the so-called Orange agent. You killed civilians and innocent people, not soldiers, innocent people every single war you went. You went to wars more than any other country in this century, and then you have the nerve to talk about killing innocent people.

"And now you have invented new ways to kill innocent people. You have so-called economic embargo which kills nobody other than children and elderly people, and which other than Iraq you have been placing the economic embargo on Cuba and other countries for over 35 years.

"The government in its summations and opening said that I was a terrorist. Yes, I am a terrorist and I am proud of it. And I support terrorism so long as it was against the United States Government and Israel, because you are more than terrorists; you are the one who invented terrorism and using it every day. You are butchers, liars and hypocrites [*sic*]" (*New York Times*, 1998).

The following is an old cliché on a terrorist: "One man's terrorist is another man's freedom fighter."

Again, from the preceding points of view, it can be seen that defining terrorism or the terrorist is not straightforward and never easy. Even the standard dictionary definition leaves us with the vagaries and ambiguities of other words typically associated with terrorism such as in the definitions of *unlawful* and *public welfare* (Sauter and Carafano, 2005).

At this point, the reader may wonder, "Why should we care; that is, what difference does it make what the definition of terrorist or terrorism is?" Definitions are important because in order to prepare for the terrorism contingency, domestic or international, we must have some feel, as with any other problem, for what it is we are dealing with. We are fighting a war of ideas. We must attempt to understand both sides of the argument, even though the terrorist's side makes no sense to an American or other freedom-loving occupants of the globe.

Finally, while it is difficult to pinpoint an exact definition of terrorism, we certainly have little difficulty in identifying it when we see it, when we feel it, or when we suffer from it. Consider, for example, the earlier account of Daniel's actions and the chlorine disaster. Put yourself in the place of those college students who were simply leaving campus buildings to cross the parking lot to their cars to make the journey home. In particular, put yourself in the place of one of those female students who was pregnant and who, as she approached her parked car, looked up and saw that yellow-green cloud of death racing with the wind toward her, eventually surrounding her, and then killing her. She could not have known that an American terrorist had, on U.S. soil, caused an act of terrorism that killed her. No, she did not know that. There is one thing she knew for certain; she knew that crushing feeling of terror as she struggled to breathe. By any other name terrorism is best summed up as that absolute feeling of Terror—Terror with a capital *T*.

VOCABULARY OF HATE

After 9/11, several authors published and the media transmitted seemingly endless accounts of various hate groups operating throughout the globe. Overnight, Americans became aware of various theories, philosophies, and terminology very few had ever heard of or thought about. This trend is ongoing and seemingly never ending.

Various pundits, so-called experts on the "new" genre of terrorism, have stated that for Americans to understand why foreign terrorists behead innocent people (or anyone else, for that matter) on television or blow up hospitals full of the sick or wounded or schoolhouses full of children, they must get inside the mind of a terrorist.

The average American might ask, "Get inside the mind of a terrorist? How do you get inside the mind of madmen?"

This is where we make our first mistake, thinking the terrorists act in the manner they do because they are mad, irrational, disturbed, or psychotic. In the case of Timothy McVeigh, we might be able to characterize him and his actions in this manner. Yet McVeigh is the exception that proves the rule—terrorist attacks, by real terrorists, are primarily planned beforehand by a group. It is important to remember that McVeigh acted primarily alone.

The terrorists that crashed airplanes into the Twin Towers, Pentagon, and the farm field in Pennsylvania were all of the same mindset; they worked as a group. Likewise, the terrorists that are presently attacking Baghdad every day work as a group. Terrorists that did all the damage in Bali and Spain and elsewhere acted as a group. Thus, though we would like to classify all the terrorists as we classify Timothy McVeigh, we can't do that. One madman working alone is something we can reasonably assume. However, thinking that hundreds or thousands of like-minded madmen all work in groups is a stretch. The cold-blooded manner in which terrorists go about their business suggests that they are not crazy, insane, or mad, but instead extremely harsh and calculating. If we dismiss them as madmen, we underestimate their intelligence. When we do that, we lose. No, we cannot underestimate the enemy—the terrorists. They are smart, cold-blooded, and calculating. In order to protect our critical water infrastructure, we must be smarter and expect the unexpected—we must be proactive and not just reactive in implementing our countermeasures. Simply, we must outsmart the enemy.

The Language of Terrorism

Anyone who is going to work at improving the security of America's critical infrastructure must be well versed in the goals and techniques used by the terrorists. Moreover, we cannot implement effective countermeasures unless we know our vulnerabilities. Along with this, we must also understand the terrorist. We must not only understand what they are capable of doing, but also have some feel for their language or vocabulary, which will help us to understand where they are coming from and where they might be headed, so to speak.

As with any other technical presentation, understanding the information presented is difficult unless a common vocabulary is established. Voltaire said it best: "If you wish to converse with me, please define your terms." It is difficult enough to understand terrorists and terrorism; thus, we must be familiar with terms they use and that are used to describe them, their techniques, and their actions.

Definition of Terms

Abu Sayyaf—Meaning "bearer of the sword," is the smaller of the two Islamist groups whose goal is to establish an Iranian-style Islamic state in Mindanao in the southern Philippines. In 1991, the group split from the Maro National Liberation Front. With ties to numerous Islamic fundamentalist groups, they finance their operations through kidnapping for ransom, extortion, piracy, and other criminal acts. It is also thought that they receive funding from al Qaeda. It is estimated that there are between 200 and 500 Abu Sayyaf terrorists, mostly recruited from high schools and colleges.

acid bomb—Crude bomb made by combining muriatic acid with aluminum strips in a two-liter soda bottle.

aerosol—A fine mist or spray, which contains minute particles.

Afghanistan—At the time of 9/11, Afghanistan was governed by the Taliban, and Osama bin Laden called it home. Amid U.S. air strikes, which began on October 7, 2001, the United States sent more than $300 million in humanitarian aide. In December 2001, Afghanistan reopened its embassy for the first time in more than 20 years.

aflatoxin—Toxin created by bacteria that grow on stored foods, especially on rice, peanuts, and cotton seeds.

agency—A division of government with a specific function or a nongovernmental organization (e.g., private contractor or business) that offers a particular kind of assistance. In the incident command system, agencies are defined as jurisdictional (having statutory responsibility for incident mitigation) or assisting and/or cooperating (providing resources and/or assistance).

air marshal—A federal marshal whose purpose is to ride commercial flights dressed in plain clothes and who is armed to prevent hijackings. Israel's use of air marshals on El Al is credited as the reason Israel has only had a single hijacking in 31 years.

The United States started using air marshals after September 11, 2001. Despite President Bush's urging, there are not enough air marshals to go around, so many flights do not have them.

airborne—Carried by or through the air.

al-Gama'a al-Islamiyya (The Islamic Group [IG])—Islamic organization that emerged spontaneously during the 1970s in Egyptian jails and later in Egyptian universities. After President Sadat released most of the Islamic prisoners from prisons in 1971, groups of militants organized themselves in groups and cells, and al-Gama'a al-Islamiyya was one of them.

al Jazeera—Satellite television station based in Qatar and broadcast throughout the Middle East. al Jazeera has often been called the CNN of the Arab world.

al Qaeda—Meaning "the base," is an international terrorist group founded in approximately 1989 and dedicated to opposing non-Islamic governments with force and violence. One of the principal goals of al Qaeda was to drive the U.S. armed forces out of the Saudi Arabian peninsula and Somalia by violence. The group is believed to be responsible for several terrorist attacks, including those on the U.S. embassy in Kenya and Tanzania, as well as the first and second World Trade Center bombings and the attack on the Pentagon.

al Tahwid—Palestinian group based in London that professes a desire to destroy both Israel and the Jewish people throughout Europe. Eleven al Tahwid were arrested in Germany in September 2005 as they were allegedly about to begin attacking that country.

alpha radiation—The least penetrating type of nuclear radiation. Alpha radiation is not considered dangerous unless particles enter the body.

American Airlines Flight 11—The Boeing 767 carrying eighty-one passengers, nine flight attendants, and two pilots that was hijacked and crashed into the north tower of the World Trade Center at 8:45 a.m. eastern standard time on September 11, 2001. Flight 11 was en route to Los Angeles from Boston.

American Airlines Flight 77—The Boeing 757 carrying fifty-eight passengers, four flight attendants, and two pilots that was hijacked and crashed into the Pentagon at 9:40 a.m. eastern standard time on September 11, 2001. Flight 77 was en route to Los Angeles from Dulles International Airport in Virginia.

ammonium nitrate-fuel oil (ANFO)—A powerful explosive made by mixing fertilizer and fuel oil. This type of bomb was used in the first World Trade Center attack as well as the Oklahoma City bombing.

analyte—The name assigned to a substance or feature that describes it in terms of its molecular composition, taxonomic nomenclature, or other characteristic.

anthrax—Often fatal infectious disease contracted from animals. Anthrax spores have a long survival period, a short incubation period, and the power to cause severe disability, making anthrax a bioweapon of choice by several nations.

antidote—A remedy to counteract the effects of poison.

antigen—A substance that stimulates an immune response by the body. The immune system recognizes such substances as foreign and produces antibodies to fight them.

antitoxin—An antibody which neutralizes a biological toxin.

Armed Islamic Group (GIA)—Algerian Islamic extremist group that aims to overthrow the secular regime in Algeria and replace it with an Islamic state. The GIA began its violent activities in early 1992 after Algiers voided the victory of the largest Islamic party, Islamic Salvation Front (FIS), in the December 1991 elections.

asymmetric threat—The use of crude or low-tech methods to attack a superior or more high-tech enemy.

Axis of Evil—Iran, Iraq, and North Korea, as named by President G. W. Bush during his State of the Union speech in 2002. President Bush named these countries the Axis of Evil because, he said, they threaten U.S. security by harboring terrorism.

Biosafety Level 1—Suitable for work involving well-characterized biological agents not known to consistently cause disease in healthy adult humans and of minimal potential hazard to lab personnel and the environment. Work is generally conducted on open benchtops using standard microbiological practices.

Biosafety Level 2—Suitable for work involving biological agents of moderate potential hazard to personnel and the environment. Lab personnel should have specific training in handling pathogenic agents and be directed by competent scientists. Access to the lab should be limited when work is being conducted, extreme precautions should be taken with contaminated sharp items, and certain procedures should be conducted in biological safety cabinets or other physical containment equipment if there is a risk of creating infectious aerosols or splashes.

Biosafety Level 3—Suitable for work done with indigenous or exotic biological agents that may cause serious or potentially lethal disease as a result of exposure by inhalation. Lab personnel must have specific training in handling pathogenic and potentially lethal agents and be supervised by competent scientists who are experienced in working with these agents. All procedures involving the manipulation of infectious material are conducted within biological safety cabinets or other physical containment devices or by personnel wearing appropriate personal protective clothing and equipment. The lab must have special engineering and design features.

Biosafety Level 4—Suitable for work with the most infectious biological agents. Access to the two Biosafety Level 4 labs in the United States is highly restricted.

Bioterrorism Act—The Public Health Security and Bioterrorism Preparedness and Response Act of 2002.

BWC—The Biological Weapons Convention, officially known as the "Convention on the Prohibition of Development, Production, and Stockpiling of Bacteriological (Biological) and Toxin Weapons and Destruction." The BWC works toward general and complete disarmament, including the prohibition and elimination of all types of weapons of mass destruction.

Baath Party—The official political party in Iraq until the United States "debaathified" Iraq in May 2003, after a war which lasted a little over a month. Saddam Hussein, the former ruler of the Baath party, was targeted by American-led coalition forces and fled. Baath party members have been officially banned from participating in any new government in Iraq.

Beltway Sniper—For nearly a month in October 2002, the Washington D.C., Maryland, and Virginia area was the hunting grounds for 41-year-old John Allen Muhammad and 17-year-old Lee Boyd Malvo. Dubbed "the Beltway Sniper" by the media, they shot people at seemingly random places such as schools, restaurants, and gas stations.

biochemical warfare—Collective term for use of both chemical warfare and biological warfare weapons.

biochemterrorism—Terrorism using biological or chemical agents as weapons.

biological ammunition—Ammunition designed specifically to release a biological agent used as the warhead for biological weapons. Biological ammunition may take many forms, such as a missile warhead or bomb.

biological attacks—The deliberate release of germs or other biological substances that cause illness.

bioterrorism—The use of biological agents in a terrorist operation. Biological toxins include anthrax, ricin, botulism, the plague, smallpox, and tularemia.

biowarfare—The use of biological agents to cause harm to targeted people either directly, by bringing the people into contact with the agents; or indirectly, by infecting other animals and plants, which would in turn cause harm to people.

blister agents—Agents which cause pain and incapacitation instead of death and might be used to injure many people at once, thereby overloading medical facilities and causing fear in the population. Mustard gas is the best-known blister agent.

blood agents—Agents based on cyanide compounds. These are more likely to be used for assassination than for terrorism.

botulism—The botulinum toxin is exceedingly lethal and quite simple to produce. It takes just a small amount of the toxin to destroy the central nervous system. Botulism may be contracted by the ingestion of contaminated food or through breaks or cuts in the skin. Food supply contamination or aerosol dissemination of the botulinum toxin are the two methods most likely to be used by terrorists.

Bush Doctrine—The policy that holds responsible nations that harbor or support terrorist organizations and says that such countries are considered hostile to the United States. President Bush described the doctrine as follows: "A country that harbors terrorists will either deliver the terrorist or share in their fate. . . . People have to choose sides. They are either with the terrorists, or they're with us."

Camp X-Ray—The Guantánamo Bay, Cuba, camp that houses al Qaeda and Taliban prisoners.

carrier—Person or animal that is potentially a source of infection by carrying an infectious agent without visible symptoms of the disease.

causative agent—The pathogen, chemical, or other substance that is the cause of disease or death in an individual.

cell—The smallest unit within a guerrilla or terrorist group. A cell generally consists of two to five people dedicated to a terrorist cause. The formation of cells is born of the concept that an apparent "leaderless resistance" makes it hard for counterterrorists to penetrate.

chain of custody—The tracking and documentation of physical control of evidence.

chemical agent—Toxic substances intended to be used for operations to debilitate, immobilize, or kill military or civilian personnel.

chemical ammunition—Munitions, commonly a missile, bomb, rocket, or artillery shell, designed to deliver chemical agents.

chemical attack—The intentional release of toxic liquid, gas or solid, in order to poison the environment or people.

chemical warfare—The use of toxic chemicals as weapons, not including herbicide, to defoliate battlegrounds or riot control agents such as gas or Mace.

chemical weapons—Weapons that produce effects on living targets via toxic chemical properties. Examples would be sarin, VX nerve gas, or mustard gas.

chemoterrorism—The use of chemical agents in a terrorist operation. Well-known chemical agents include sarin and VX nerve gas.

choking agent—Compounds that primarily injure the respiratory tract (i.e., nose, throat, and lungs). In extreme cases membranes swell up, lungs become filled with liquid, and death results from lack of oxygen.

Cipro—Bayer's antibiotic which combats inhalation anthrax.

confirmed—In the context of the threat evaluation process, a water contamination incident is definitive evidence that the water has been contaminated.

counterterrorism—Measures used to prevent, preempt, or retaliate against terrorist attacks.

credible—In the context of the threat evaluation process, a water contamination threat is characterized as "credible" if information collected during the threat evaluation process corroborates information from the threat warning.

cutaneous—Related to or entering through the skin.

cutaneous anthrax—Contracted via broken skin. The infection spreads through the bloodstream causing cyanosis, shock, sweating, and finally death.

cyanide agents—Used by Iraq in the Iran war and against the Kurds in the 1980s and also by the Nazis in the gas chambers of concentration camps, cyanide agents are a colorless liquid which is inhaled in its gaseous form while liquid cyanide and cyanide salts are absorbed by the skin. Symptoms are headache, palpitations, dizziness, and respiratory problems followed later by vomiting, convulsions, respiratory failure, unconsciousness, and eventually death.

cyberterrorism—Attacks on computer networks or systems, generally by hackers working with or for terrorist groups. Some forms of cyberterrorism include denial of service attacks, inserting viruses, or stealing data.

dirty bomb—A makeshift nuclear device that is created from radioactive nuclear waste material. While not a nuclear blast, an explosion of a dirty bomb causes localized radioactive contamination as the nuclear waste material is carried into the atmosphere where it is dispersed by the wind.

Ebola—Ebola hemorrhagic fever (EHF) is a severe, often-fatal disease in nonhuman primates, such as monkeys, chimpanzees, and gorillas, and in humans. Ebola has appeared sporadically since 1976, when it was first recognized.

eBomb (or e-bomb)—Electromagnetic bomb that produces a brief pulse of energy, which affects electronic circuitry. At low levels, the pulse temporarily disables electronics systems, including computers, radios, and transportation systems. High levels completely destroy circuitry, causing mass disruption of infrastructure while sparing life and property.

ecoterrorism—Sabotage intended to hinder activities that are considered damaging to the environment.

Euroterrorism—Associated with left-wing terrorism of the 1960s, 1970s, and1980s involving the Red Brigade, Red Army Faction, and November 17th Group, among other groups that targeted American interests in Europe and NATO. Other groups include Orange Volunteers, Red Hand Defenders, Continuity IRA, Loyalist Volunteer Force, Ulster Defense Association, and First of October Anti-Fascist Resistance Group.

fallout—The descent to the earth's surface of particles contaminated with radioactive material from a radioactive cloud. The term can also be applied to the contaminated particulate matter itself.

Fatah—Meaning "conquest by means of jihad," the political organization created in the 1960s and led by Yasser Arafat. With both a military and an intelligence wing, it has carried out terrorist attacks on Israel since 1965. It joined the Palestinian Liberation Organization (PLO) in 1968. Since 9/11, the Fatah was blamed for attempting to smuggle 50 tons of weapons into Israel.

fatwa—A legal ruling regarding Islamic law.

Fedayeen Saddam—Iraq's former paramilitary organization said to be an equivalent to the Nazi's SS. The militia was loyal to Saddam Hussein and was responsible for using brutality on civilians who were not loyal to the policies of Saddam. They did not dress in uniform.

filtrate—In ultrafiltration, the water that passes through the membrane and contains only particles smaller than the molecular weight cutoff of the membrane.

Frustration-Aggression Hypothesis—A hypothesis that every frustration leads to some form of aggression and every aggressive act results from some prior frustration.

As defined by Gurr (1968): "The necessary precondition for violent civil conflict is relative deprivation, defined as actors' perception of discrepancy between their value expectations and their environment's apparent value capabilities. This deprivation may be individual or collective."

fundamentalism—Conservative religious authoritarianism. Fundamentalism is not specific to Islam; it exists in all faiths. Characteristics include literal interpretation of scriptures and a strict adherence to traditional doctrines and practices.

Geneva Protocol 1925—The first treaty to prohibit the use of biological weapons, also known as the Protocol for the Prohibition of the Use in War of Asphyxiating, Poisonous or Other Gases and Bacteriological Methods of Warfare.

germ warfare—The use of biological agents to cause harm to targeted people either directly, by bringing the people into contact with the agents; or indirectly, by infecting other animals and plants, which would in turn cause harm to people.

glanders—An infectious bacterial disease known to cause inflammation in horses, donkeys, mules, goats, dogs, and cats. Human infection has not been seen since 1945, but because so few organisms are required to cause the disease, it is considered a potential agent for biological warfare.

grab sample—A single sample collected at a particular time and place that represents the composition of the water, air, or soil only at that time and location.

ground zero—From 1946 until 9/11, ground zero was the point directly above, below, or at which a nuclear explosion occurs or the center or origin of rapid, intense, or violent activity or change. After 9/11, the term, when used with initial capital letters, refers to the ground at the epicenter of the World Trade Center attacks.

guerrilla warfare—The term was invented to describe the tactics Spain used to resist Napoleon, though the tactic itself has been around much longer. Literally, it means "little war." Guerrilla warfare features cells and utilizes no front line. The oldest form of asymmetric warfare, guerrilla warfare is based on sabotage and ambush with the objective of destabilizing the government through lengthy and low-intensity confrontation.

Hamas—A radical Islamic organization that operates primarily in the West Bank and Gaza Strip whose goal is to establish an Islamic Palestinian state in place of Israel. On the one hand, Hamas operates overtly in its capacity as a social services deliverer, but its activists have also conducted many attacks, including suicide bombings, against Israeli civilians and military targets.

hazard assessment—The process of evaluating available information about the site to identify potential hazards that might pose a risk to the site characterization team. The hazard assessment results in assigning one of four levels to risk: lower hazard, radiological hazard, high chemical hazard, or high biological hazard.

hemorrhagic fevers—In general, the term *viral hemorrhagic* fever is used to describe a severe multisystem syndrome wherein the overall vascular system is damaged, and the body becomes unable to regulate itself. These symptoms are often accompanied by hemorrhage; however, the bleeding itself is not usually life threatening. Some types of hemorrhagic fever viruses can cause relatively mild illnesses.

Hizbollah (Hezbollah)—Meaning, "the Party of God," one of many terrorist organizations that seek the destruction of Israel and the United States. They have taken credit for numerous bombings against civilians and have declared that civilian targets are warranted. Hezbollah claims that it sees no legitimacy for the existence of Israel and that their conflict becomes one of legitimacy that is based on religious ideals.

Homeland Security Office—Agency organized after 9/11, with former Pennsylvania governor Tom Ridge heading it up. The Office of Homeland Security is at the top of approximately 40 federal agencies charged with protecting the United States against terrorism.

homicide bombings—The White House coined the term to replace the old "suicide bombings."

incident—A confirmed occurrence that requires response actions to prevent or minimize loss of life or damage to property and/or natural resources. A drinking water contamination incident occurs when the presence of a harmful contaminant has been confirmed.

inhalation anthrax—Contracted by inhaling anthrax spores. This results in pneumonia, sometimes meningitis, and finally death.

intifada (intifadah or Intifadah, from Arabic "shaking off")—The two intifadas are similar in that both were originally characterized by civil disobedience by the Palestinians that escalated into the use of terror. In 1987, following the killing of several Arabs in the Gaza Strip, the first intifada began and went on until 1993. The second intifada began in September 2000, following Ariel Sharon's visit to the Temple Mount.

Islam—Meaning, "submit," the faith practiced by followers of Muhammad. Islam claims to have more than a billion believers worldwide.

jihad—Meaning, "struggle," but the definition is a subject of vast debate. There are two definitions generally accepted. The first is a struggle against oppression, whether political or religious. The second is the struggle within oneself, or a spiritual struggle.

LD50—A dose of a substance which kills 50 percent of those infected.

Laboratory Response Network (LRN)—A network of labs developed by the Centers for Disease Control, Association of Public Health Laboratories, and FBI for the express purpose of dealing with bioterrorism threats, including pathogens and some biotoxins.

Lassa fever—An acute, often fatal, viral disease characterized by high fever, ulcers of the mucous membranes, headaches, and disturbances of the gastrointestinal system.

kneecapping—This common punishment used by Northern Ireland's IRA, which involves collaborating with the British.

Koran—The holy book of Islam, considered by Muslims to contain the revelations of God to Mohammed. It is also called the Qu'ran.

mindset—A noun defined by *American Heritage Dictionary* as "1. A fixed mental attitude or disposition that predetermines a person's response to and interpretation of situations; 2. an inclination or a habit." *Merriam Webster's Collegiate Dictionary* (10th ed.) defines it as "1. A mental attitude or inclination; 2. a fixed state of mind." The term dates from 1926 but apparently is not included in dictionaries of psychology.

molotov cocktail—A crude incendiary bomb made of a bottle filled with flammable liquid and fitted with a rag wick.

monkeypox—The Russian bioweapon program worked with this virus, which is in the same family as smallpox. In June 2003, a spate of human monkeypox cases was reported in the American Midwest. This was the first time that monkeypox was seen in North America, and it was the first time that monkeypox was transferred from animal to human. There was some speculation that it was a bioattack.

mullah—A Muslim, usually holding an official post, who is trained in traditional religious doctrine and law and doctrine.

Muslim (also Moslem)—Followers of the teachings of Mohammed, or Islam.

mustard gas—Blistering agents that cause severe damage to the eyes, internal organs, and respiratory system. Produced for the first time in 1822, mustard gas was not used until World War I. Victims suffer the effects of mustard gas 30–40 years after exposure.

narcoterrorism—The view of many counterterrorist experts that there exists an alliance between drug traffickers and political terrorists.

National Pharmaceutical Stockpile—A stock of vaccines and antidotes that are stored at the Centers for Disease Control in Atlanta, to be used in the event of biological warfare.

nerve agent—The Nazis used the first nerve agents, which were insecticides developed into chemical weapons. Some of the better-known nerve agents include VX, sarin, soman, and tabun. These agents are used because only a small quantity is necessary to inflict a substantial damage. Nerve agents can be inhaled or can absorb through intact skin.

nuclear blast—An explosion of any nuclear material that is accompanied by a pressure wave, intense light and heat, and widespread radioactive fallout, which can contaminate the air, water, and ground surface for miles around.

opportunity contaminant—Contaminants that might be readily available in a particular area, even through they may not be highly toxic, infectious, or easily dispersed and stable in treated drinking water.

Osama bin Laden (also spelled "Usama")—A native of Saudi Arabia, was born the 17th of 24 sons of Saudi Arabian builder Mohammed bin Oud bin Laden, a Yemeni immigrant. Early in his career, he helped the mujahedeen fight the Soviet Union by recruiting Arabs and building facilities. He hates the United States, and apparently this is because he views the United States as having desecrated holy ground in Saudi Arabia with its presence during the first Gulf War. Expelled from Saudi Arabia in 1991 and from Sudan in 1996, he operated terrorist training camps in Afghanistan. His global network al Qaeda is credited with the attacks on the United States on September 11, 2001, the attack on the USS *Cole* in 2000, and a number of other terrorist attacks.

pathogen—Any agent that can cause disease.

plague—The pneumonic plague, which is more likely to be used in connection with terrorism, is naturally carried by rodents and fleas but can be aerosolized and sprayed from crop dusters. A 1970 World Health Organization assessment asserted that, in a worst case scenario, a dissemination of 50 kilograms in an aerosol over a city of five million people could result in 150,000 cases of pneumonic plague, 80,000–100,000 of which would require hospitalization, and 36,000 of which would be presumed fatal.

political terrorism—Terrorist acts directed at governments and their agents and motivated by political goals (i.e., national liberation).

"possible"—In the context of the threat evaluation process, a water contamination threat is characterized as "possible" if the circumstances of the threat warning appear to have provided an opportunity for contamination.

potassium iodide—Food and Drug Administration–approved nonprescription drug used as a blocking agent to prevent the thyroid gland from absorbing radioactive iodine.

presumptive results—Results of chemical and/or biological field testing that need to be confirmed by further lab analysis. Typically used in reference to the analysis of pathogens.

psychopath—A mentally ill or unstable person, especially one having a psychopathic personality (q.v.), according to *Webster's*.

psychopathology—The study of psychological and behavioral dysfunction occurring in mental disorder or in social disorganization, according to *Webster's*.

psychopathy—A mental disorder, especially an extreme mental disorder usually marked by egocentric and antisocial activity, according to *Webster's*.

psychotic—Of, relating to, or affected with psychosis, which is a fundamental mental derangement (as schizophrenia) characterized by defective or lost contact with reality, according to *Webster's*.

rapid field testing—Analysis of water during site characterization using rapid field water testing technology in an attempt to tentatively identify contaminants or unusual water quality.

retentate—In ultrafiltration, the retentate is the solution that contains the particles that do not pass through the membrane filter. The retentate is also called the concentrate.

ricin—A stable toxin easily made from the mash that remains after processing castor beans. At one time, it was used as an oral laxative, castor oil; castor oil causes diarrhea, nausea, vomiting, abdominal cramps, internal bleeding, liver and kidney failure, and circulatory failure. There is not an antidote.

salmonella—An infection caused by a gram-negative bacillus, a germ of the *Salmonella* genus. Infection with this bacteria may involve only the intestinal tract or may be spread from the intestines to the bloodstream and then to other sites in the body.

Symptoms of salmonella enteritis include diarrhea, nausea, fever, abdominal pain, and fever. Dehydration resulting from the diarrhea can cause death, and the disease can cause meningitis or septicemia. The incubation period is between eight and 48 hours, while the acute period of the illness can last for one to two weeks.

sarin—Colorless, odorless gas. With a lethal dose of .5 milligrams (a pinpoint-sized droplet), it is 26 times more deadly than cyanide gas. Because the vapor is heavier than air, it hovers close to the ground. Sarin degrades quickly in humid weather, but sarin's life expectancy increases as temperature gets higher, regardless of how humid it is.

Sentinel Laboratory—A Laboratory Response Network (LRN) lab that reports unusual results that might indicate a possible outbreak and refers specimens that may contain select biological agents to reference labs within the LRN.

site characterization—The process of collecting information from an investigation site in order to support the evaluation of a drinking water contamination threat. Site characterization activities include the site investigation, field safety screening, rapid field testing of the water, and sample collection.

sleeper cell—A small cell that keeps itself undetected until it can "awaken" and cause havoc.

smallpox—The first biological weapon, used during the eighteenth century, smallpox killed 300 million people in the nineteenth century. There is no specific treatment for smallpox disease, and the only prevention is vaccination. This currently poses a problem, since the vaccine was discontinued in 1970 after the World Health Organization declared smallpox eradicated. Incubation is seven to 17 days, during which time the carrier is not contagious. Thirty percent of people exposed are infected, and the disease has a 30 percent mortality rate.

sociopath—Basically synonymous with psychopath (s.v.). Sociopathic symptoms in the adult sociopath include an inability to tolerate delay or frustration, a lack of guilty feelings, a relative lack of anxiety, a lack of compassion for others, a hypersensitivity to personal ills, and a lack of responsibility. Many authors prefer the term *sociopath* because this type of person had defective socialization and a deficient childhood.

sociopathic—Of, relating to, or characterized by asocial or antisocial behavior or a psychopathic (s.v.) personality, accord to *Webster's*.

spore—An asexual, usually single-celled reproductive body of plants such as fungi, mosses, or ferns; a microorganism, as a bacterium, in a resting or dormant state.

terrorist group—A group that practices or has significant elements that are involved in terrorism.

threat—An indication that a harmful incident, such as contamination of the drinking water supply, may have occurred. The threat may be direct, such as a verbal or written threat, or circumstantial, such as a security breach or unusual water quality.

toxin—Poisonous substance, which is produced by living organisms, capable of causing disease when introduced into the body tissues.

transponder—A device on an airliner which sends out a signal allowing air traffic controllers to track an airplane. Transponders were disabled in some of the planes highjacked on 9/11.

Transportation Security Administration (TSA)—A new agency created by the Patriot Act of 2001 for the purpose of overseeing technology and security in American airports.

tularemia—Tularemia is an infectious disease caused by the hardy bacterium *Francisella tularensis*, found in animals, especially rabbits, hares, and rodents. Symptoms depend on how the person was exposed to tularemia but can include difficulty breathing, chest pain, bloody sputum, swollen and painful lymph glands, ulcers on the mouth or skin, swollen and painful eyes, and sore throat. Symptoms usually appear three to five days after exposures but sometimes will take up to two weeks. Tularemia is not spread from person to person, so people who have it need not be isolated.

ultrafiltration—A filtration process for water that uses membranes to preferentially separate very small particles that are larger than the membrane's molecular weight cutoff, typically greater than 10,000 daltons. (A dalton is a unit of mass, defined as one-twelfth the mass of a carbon-12 nucleus. It's also called the atomic mass unit, abbreviated as either *amu* or *u*.)

vector—An organism that carries germs from one host to another.

vesicle—A blister filled with fluid.

weapons of mass destruction (WMD)—According to the National Defense Authorization Act, WMDs are any weapon or device that is intended, or has the capability, to cause death or serious bodily injury to a significant number of people through the release, dissemination, or impact of the following:

- toxic or poisonous chemicals or their precursors
- a disease organism
- radiation or radioactivity

xenophobia—Irrational fear of strangers or those who are different from oneself.

zyklon b—A form of hydrogen cyanide. Symptoms of inhalation include increased respiratory rate, restlessness, headache, and giddiness followed later by convulsions, vomiting, respiratory failure, and unconsciousness. Used in the Nazi gas chambers in World War II.

REFERENCES

Bergen, P. L. 2002. *Holy War, Inc: Inside the Secret World of Osama bin Ladin*, 21–22. New York: Touchstone Press.

FBI. 2006. *World Conflict Quarterly*, at www.globalterrorism.o1.com.

Gurr, T. R. 1968. Psychological factors in civil violence. *World Politics* 20, no. 32 (January): 245–78.

Henry, K. 2002. New face of security. *Government Security* (April): 30–37.

National Strategy for Homeland Security. 2006. The White House, at www.whitehouse/homeland.

New York Times. 1998. Excerpt from court testimony. January 9.

Sauter, M. A., and J. J. Carafano. 2005. *Homeland Security: A Complete Guide to Understanding, Preventing, and Surviving Terrorism*. New York: McGraw-Hill.

Spellman, F. R. 1997. *A Guide to Compliance for Process Safety Management/Risk Management Planning (PSM/RMP)*. Lancaster, PA: Technomic Publishing Company.

Spellman, F. R. 2000, 2002. Violence in the workplace: Security Concerns. From a series of lectures presented to environmental health students at Old Dominion University, Norfolk, VA.

U.S. Congress. 2005. Annual country reports on terrorism. 22 USC, Chapter 38, Section 2656f.

2

Water/Wastewater Infrastructure

Those physical and cyber-based systems essential to the minimum operations of the economy and government.

—*Definition of critical infrastructure in President Bill Clinton's Presidential Decision Directive, No. 63*

As part of the 1998 presidential directive, quoted above, the federal government began to assess the vulnerability of U.S. infrastructure and develop ways to protect it. The federal government identified several categories as critical infrastructure: aviation, highways and mass transit; pipelines, rail, and waterborne commerce; public health services; electric power; oil and gas production and storage; information and communications; banking and finance; and the drinking water supply. Notice that wastewater treatment was not initially listed. However, after 9/11, utility infrastructure experts (in terms of securing their premises and preparing for emergency response) pointed out the need for wastewater treatment systems to implement many of the same security measures as water treatment facilities.

In response to 9/11, the Environmental Protection Agency (EPA) increased its efforts in defending the nation's water infrastructure against a domestic and/or international terrorist attack. A major step in this effort was the establishment of the Water Protection Task Force in October 2001. The EPA Task Force includes experts in a variety of subjects, including drinking water and wastewater treatment, security, training and outreach, and funding. The goal of the task force is to help make drinking water and wastewater infrastructure as safe as possible, as quickly as possible. Working with the states, tribes, utilities, and other appropriate partners, the EPA strives to provide utilities with the best information and tools available to reduce their vulnerability to terrorist attacks. As might be imagined, this effort is ongoing and is being pursued on an accelerated schedule.

According to the EPA (2002), the Water Protection Task Force works with approximately 168,000 public water systems in the United States; 54,000 of these waters systems are community systems that supply water to 264 million consumers. Approximately 80 percent of the population in the United States is served by only 7 percent of the systems—large utilities that serve more than 10,000 consumers each. Most systems, conversely, are small and serve relatively small populations. In regards to wastewater, 20 percent of the approximately 16,000 publicly owned treatment works (POTWs) serve the major metropolitan areas and consequently a large portion of the population. The remaining population is served by privately owned utilities or by onsite systems, such as septic tanks. POTWs discharge treated effluent into receiving waters and are regulated under the Clean Water Act.

WATER

When color photographs of the earth as it appears from space were first published, it was a revelation: They showed our planet to be astonishingly beautiful. We were taken by surprise. What makes the earth so beautiful is its abundant water. The great expanses of vivid, blue ocean with swirling, sunlit clouds above them should not have caused surprise, but the reality exceeded everybody's expectations. The pictures must have brought home to all who saw them the importance of water to our planet

—*E. C. Pielou, 1998*

Water is a contradiction, a riddle.

How?

Consider the Chinese proverb that states, "Water can both float and sink a boat."

Water's presence everywhere feeds these contradictions. S. A. Lewis (1996) points out that "water is the key ingredient of mother's milk and snake venom, honey and tears" (p. 90).

Leonardo da Vinci gave us insight into more of water's apparent contradictions:

Water is sometimes sharp and sometimes strong, sometimes acid and sometimes bitter;

Water is sometimes sweet and sometimes thick or thin;

Water sometimes brings hurt or pestilence, sometimes health-giving, sometimes poisonous.

Water suffers changes into as many natures as are the different places through which it passes.

Water, as with the mirror that changes with the color of its object, so it alters with the nature of the place, becoming: noisome, laxative, astringent, sulfurous, salt, incarnadined, mournful, raging, angry, red, yellow, green, black, blue, greasy, fat or slim.

Water sometimes starts a conflagration, sometimes it extinguishes one.

Water is warm and is cold.

Water carries away or sets down.
Water hollows out or builds up.
Water tears down or establishes.
Water empties or fills.
Water raises itself or burrows down.
Water spreads or is still.
Water is the cause at times of life or death, or increase of privation, nourishes at times and at others does the contrary.
Water, at times has a tang, at times it is without savor.
Water sometimes submerges the valleys with great flood.
In time and with water, everything changes.

We can sum up water's contradictions by simply stating that though the globe is awash in it, water is no single thing, but an elemental force that shapes our existence. Da Vinci's last contradiction, "In time and with water, everything changes," concerns us most.

Why?

Because next to the air we breathe, the water we drink is most important to us—to all of us. Water is no less important than our air, simply less urgent. Simply stated, for us, for all of us, though we treat it casually, unthinkingly, water is not a novelty, but a necessity—we simply cannot live without water.

Some might view these statements about the vital importance of water (commonly incorrectly viewed as a rather plain and simple substance) as nothing more than hyperbole, exaggeration, panic, or overstatement. But are they? Is our concern over safe drinking water really an exaggeration? I think not. Why? Because we absolutely know and understand, for example, this simple point: We were born of water, and to live, we must be sustained by it.

Development and protection of safe drinking water supplies is a major concern today. This may seem strange to the average person (depending, of course, where this individual resides) who might literally be surrounded by various water bodies (we are literally surrounded by water, aren't we?). However, drinking water practitioners (those responsible for finding a source, certifying its safety, and providing it to the consumer) know better. Moreover, post-9/11, those responsible for maintaining the safety/security of our drinking water supply know better.

Prior to 9/11, the drinking water practitioner knew, for example, that two key concerns drove the development of safe drinking water supplies: quantity and quality (Q and Q). Herein was the problem. Quantity may indeed be a major issue (a limiting factor) for a particular location—often the case simply because water suitable for consumption is not evenly distributed throughout the world. Those locations fortunate to have an ample supply of surface water or groundwater may not have a quantity problem, as long as the quantity is large enough to fulfill the needs of all its consumers. But again, not every geographical location is fortunate enough to have an adequate water supply—that is, the quantity of water available to satisfy residents' needs. This is one of the primary reasons, of course, that major portions of the globe are either uninhabited or sparsely populated.

The other key concern was/is water quality. Obviously, having a sufficient quantity of freshwater available does little good if the water is unsafe for consumption or for other uses.

A new key concern, because of 9/11, is water safety/security, of course. We can say that the equation Q and Q is been augmented with an S, making the new equation Q and Q and S. When you combine water safety with water security, the S in this equation morphs to a super-sized **S**. Protecting the nation's water supply is no small task; it is no smaller or any less of a concern than protecting our borders, our buildings, and our people from terrorists.

The bottom line is this: Arguably, no human, scientific, or technological development has impacted humankind's quality of life and lifespan in general more than the purification of drinking water. Thus, in the present climate, we must be super vigilant in protecting this critical, life-sustaining resource from harm.

In order to understand what critical water infrastructure is, it is important for the nonengineering professional to have some basic understanding what that infrastructure comprises. Therefore, in the following sections, a brief overview of water infrastructure is provided.

Water Purification

Water treatment brings raw water up to drinking water quality. The processes this entails depend on the quality of the water source. Surface water sources (lakes, rivers, reservoirs, and impoundments) generally require higher levels of treatment than groundwater sources. Groundwater sources may incur higher operating costs from machinery but may require only simple disinfection.

Water Treatment Unit Processes

Treatment for raw water taken from groundwater and surface water supplies differs somewhat, but one commonly employed treatment technology illustrates many of the unit processes involved. Primary treatment processes for surface water supplies include the basic water treatment processes, shown in table 2.1, and hardness removal (not mandatory), as well as the following:

- Intake to bring in the best possible quality the source can provide for treatment
- Screening to remove floating and suspended debris of a certain size

- Chemical mixing with the water to allow suspended solids to coagulate into larger particles that settle more easily
- Coagulation, a chemical water treatment method that causes small particles to stick together to form larger particles
- Flocculation to gently mix the coagulant and water, encouraging large floc particle formation
- Sedimentation to slow the flow so that gravity settles the floc
- Sludge processing to remove the solids and liquids collected in the settling tank, and to dewater and dispose of them
- Disinfection to ensure the water contains no harmful pathogens

Once water from the source has entered the plant as influent, water treatment processes break down into two parts. The first part, clarification, consists of screening, coagulation, flocculation, sedimentation, and filtration. Clarification processes go far in potable water production, but while they do remove many microorganisms from the raw water, they cannot produce water free of microbial pathogens. The final step, disinfection, destroys or inactivates disease-causing infection agents (see table 2.1).

Table 2.1 Basic Water Treatment Processes

Process/Step	*Purpose*
Intake	Conveys water from source to treatment plant
Screening	Removes large debris (leaves, sticks, fish) that could foul or damage plant equipment
Chemical pretreatment	Conditions the water for removing algae and other aquatic nuisances
Presedimentation	Removes gravel, sand, silt, and other gritty materials
Microstraining	Removes algae, aquatic plants, and remaining debris
Chemical feed and rapid mix	Adds chemicals—coagulants, pH adjusters, etc.
Coagulation/flocculation	Converts nonsettleable or settleable particles
Sedimentation	Removes settleable particles
Softening	Removes hardness-causing chemicals
Filtration	Removes particles of solid matter—includes biological contamination and turbidity
Disinfection	Kills disease-causing organisms
Adsorption using granular activated carbon (GAC)	Removes radon and many organic chemicals, including pesticides, solvents, and trihalomethanes (THMs)
Aeration	Removes volatile organic chemicals (VOCs), radon, H2S, and other dissolved gases; oxidizes iron and manganese
Corrosion control	Prevents scaling and corrosion
Reverse osmosis, electrodialysis	Removes nearly all inorganic contaminants
Ion exchange	Removes some inorganic contaminants including hardness-causing chemicals
Activated alumina	Removes some inorganic contamination
Oxidation filtration	Removes some inorganic contaminants—iron, manganese, radium, etc.

Source: Adapted from AWWA, *Introduction to Water Treatment*, Vol. 2, 1984.

Water Distribution

A municipality's water distribution and conveyance system serves two purposes: It carries raw water from a source to the plant, and it carries the finished potable water to the consumer. Water distribution systems consist of seven basic elements:

- Sources (wells or surface water)
- Storage facilities (reservoirs)
- Transmission facilities for influent
- Treatment facilities
- Transmission facilities for effluent
- Intermediate points (standpipes or water towers)
- Distribution facilities

Distribution Systems

Gravity distribution, pumping without storage, or pumping with storage are the three common water distribution methods in use. When the water supply source is well above the community's elevation, gravity distribution is possible. The least desirable method, pumping without storage, provides no reserve flow and pressures fluctuate substantially. With this method, facilities must use sophisticated control systems to meet unpredictable demand. Pumping with storage is the most common method of distribution (McGhee, 1991).

In pumping with storage, water in typical community water supply systems is carried under pressure (pumping water up into tanks that store water at higher elevations than the households they serve provides water pressure) through a network of buried pipes. Street mains carry the water from standpipes or water towers to service individual business, industrial, commercial, or residential needs. Mains usually have a minimum diameter of six to eight inches for adequate flows to supply buildings and for fire fighting. Pipes connected to buildings can be as small as one inch for small residences. House service lines (smaller pipes) from the main water lines transport water from the distribution network to households, where gravity's force moves the water into homes when household taps open. A primary goal for any water treatment facility is to provide enough water to meet system demands consistently and at adequate pressures.

Distribution systems generally follow street patterns. The location of treatment facilities and storage works affects distribution, as do the types of residential, commercial, and industrial development present, as well as topography. Distribution systems commonly set up zones related to different ground elevations and service pressures. Water mains are generally designed in enclosed loops to supply water to any point from at least two directions.

Distribution systems are categorized as grid systems, branching systems, or dead-end systems. Grid systems are generally considered the best distribution system. The looped and interconnected arterials and secondary mains eliminate dead ends, and allow free water circulation so that a heavy discharge from one main allows drawing water from other pipes. Branching systems supply to any point from at least two directions and include several terminals or dead ends. In new distribution systems, antiquated dead-end systems are completely avoided. Older systems with terminals often incorporate proper looping during retrofitting.

Water Distribution Process Equipment

Surface and groundwater water supply systems both generally involve canals, pipes, or other conveyances; pumping plants; distribution reservoirs or tanks to help balance water supply and demand and to control pressures; other appurtenances; and treatment works.

Storage tanks for potable water distribution come in a variety of types. Whatever the type, however, the tank interior must be properly protected and preserved from corrosion. Poor physical and material tank conditions degrade the stored water. Any tank coating or preservative that will be in contact with potable water must meet the National Sanitation Foundation Standard 61.

Tank types include the following:

Clear wells: These are used for storing filtered water from a treatment works and are also used as chlorine contact tanks.

Elevated tanks: Primarily used for maintaining an adequate and fairly uniform pressure to the service zone, elevated tanks are located above the service zone.

Stand pipes: These tanks stand on the ground and have a height greater than their diameter.

Ground-level reservoirs: These maintain the required pressures when located above the service area.

Hydropneumatic or pressure tanks: Often used in small water systems (with a well or booster pump), these tanks maintain water pressures in the system and control well pump or booster pump operation.

Surge tanks: Surge tanks are used mainly to control water hammer or to regulate water flow rather than as storage facilities.

Water stored in potable water storage facilities must be routinely properly monitored to detect problems in taste and odor, turbidity, color, and coliform present. Monitoring includes determining chlorine residual levels, turbidity, color, coliform analysis, decimal dilution, Most Probable Number analysis, and taste and odor analysis.

WASTEWATER

Once a water treatment plant or the homeowner/industry well has delivered potable water to a consumer and the consumer begins to use it, whether that consumer is within a household, business, or industry, the water begins the journey that leads it to the other side of water treatment—wastewater treatment.

Wastewater treatment takes effluent from water users as influent to wastewater treatment facilities. The wastestream is treated in a series of steps (unit processes, some similar to those used in treating raw water, and others that are more involved) and then discharged (outfalled) to a receiving body, usually a river or stream.

Wastewater treatment takes the wastes and water that compose the wastestream and restores the wastewater to its original quality. Wastewater treatment's goal is to treat the wastestream to the level that it is harmless to the receiving body. Most facilities actually set their goal higher: to treat the wastestream to achieve a water of a higher quality than the water contained in the receiving water body.

Wastewater Sources and General Constituents

Wastewater is the flow of used water (spent water) from a community and includes household wastes, commercial and industrial wastestream flows, stormwater, and groundwater. By weight, wastewater is generally only about .06 percent solids—dissolved or suspended materials carried in the 99.94 percent water flow. This extreme ratio of water to solids is essential to transport solids though the collection system.

The solids found in wastewater rarely contain only what most people consider "sewage"—human wastes. In fact, the dissolved and suspended solids that sewage can contain vary widely from community to community and are dependent, of course, on what industrial and commercial facilities contribute inflow to the treatment system. These influents mix with the more predictable residential flows and provide so many possible substances and microorganisms to wastewater that complete identification of every wastewater constituent is not only rarely possible, but also rarely a necessary undertaking.

This is not to say that the solids an area's wastewater contains cannot be predicted in a general way. In most communities, wastewater enters the wastewater treatment system through several ways, each with their own usual characteristic solids loads. Common industry practice puts these constituents into several general categories:

1. **Human and animal wastes:** Generally thought the most dangerous wastewater constituent from a human health viewpoint, domestic wastewater contains the solid and liquid discharges of humans and animals. These contribute millions of bacteria, viruses, and other organisms (some pathogenic) to the wastewater flow.
2. **Household wastes:** Domestic or residential wastewater flows may also contain paper, household cleaners, detergents, trash, garbage, and any other substance that a typical homeowner may pour or flush into the sewer system.
3. **Industrial wastes:** The materials that could be discharged from industrial processes into a collection system include chemicals, dyes, acids, alkalis, grit, detergents, and highly toxic materials. Individual industries present highly individual wastestreams, and these industry-specific characteristics depend on the industry processes used. Most of the time, industrial wastewaters can be treated within public treatment facilities without incident, but often industries must provide some level of treatment prior to their wastestream entering a public treatment system. This prevents compliance problems for the treatment facility. An industry may also choose to provide pretreatment because their own onsite treatment is more economical than paying municipality fees for advanced treatment.
4. **Stormwater runoff:** In collection systems that carry both community wastes and stormwater runoff, during and after storms, wastewater may contain large amounts of sand, gravel, road salt, and other grit as well as flood levels of water. Many communities install separate collection systems for stormwater runoff; in which cases that influent should contain grit and street debris but no domestic or sanitary wastes.

Wastewater systems vary by size and other factors, but all include a collection system and a treatment facility.

Wastewater Treatment

Wastewater must be collected, conveyed to a treatment facility, and treated to remove pollutants to a level of compliance with the National Pollutant Discharge Elimination System (NPDES) before a municipal or industrial facility can discharge it into receiving water.

The most common systems in wastewater treatment (as well as in water treatment) employ processes that combine physical, chemical, and biological methods. Wastewater treatment plants are usually classified as providing primary, secondary, or tertiary (or advanced) treatment, depending on the purification level to which they treat (see table 2.2).

Table 2.2 Wastewater Treatment Processes

Process/Step	*Purpose*
Primary treatment	Removes 90%–95% settleable solids, 40%–60% total suspended solids, and 25%–35% BOD5
Collection	Conveys wastewater from source to treatment plant
Screening	Removes debris that could foul or damage plant equipment
Shredding	Screening alternative that reduces solids to a size the plant equipment can handle
Grit removal	Removes gravel, sand, silt, and other gritty materials
Flow measurement	Provides compliance report data and treatment process information for hydraulic and organic loading calculations
Preaeration	Freshens septic wastes, reduces odors and corrosion, and improves solids separation and settling
Chemical addition	Reduces odors, neutralizes acids or bases, reduces corrosion, reduces BOD5, improves solids and grease removal, reduces loading on the plant, and aids subsequent processes
Flow equalization	Reduces or removes the wide swings in flow rates for plant loadings
Primary sedimentation	Concentrates and removes settleable organic and floatable solids from wastewater
Secondary treatment	Produces effluent with not more than 30 mg/L BOD5 and 30 mg/L suspended solids
Biological treatment	Provides BOD removal beyond that achievable by primary treatment using biological processes to convert dissolved, suspended, and colloidal organic wastes to more stable solids
Secondary sedimentation	Removes the accumulated biomass that remains after secondary treatment
Tertiary or advanced treatment	Removes pollutants, including nitrogen, phosphorus, soluble COD, and heavy metals to meet discharge or reuse criteria with respect to specific parameters
Effluent polishing	Filtration or microstraining to remove additional BOD or TSS
Nitrogen removal	Removes nutrients to help control algal blooms in the receiving body
Phosphorus removal	Removes limiting nutrients that could affect the receiving body
Land application	Controlled land application used as an effective alternative to tertiary treatment methods. Reduces TSS, BOD, phosphorous and nitrogen compounds, as well as refractory organics
Disinfection	Destroys any pathogens in the effluent that survived treatment
Dechlorination	Protects aquatic life from high chlorine concentrations, needed to comply with various regulations
Discharge	Releases treated effluent back to the environment through evaporation, direct discharge, or beneficial reuse.
Solids treatment	Transforms sludge to biosolids for use as soil conditioners or amendments

In primary treatment plants, physical processes (screening and sedimentation) remove a portion of the pollutants that will settle or float. Pollutants too large to pass through simple screening devices are also removed, followed by disinfection. Primary treatment typically removes about 35 percent of the biochemical oxygen demand (BOD) and 60 percent of the suspended solids.

Secondary treatment plants use the physical processes employed by primary treatment but augment the processes with the microbial oxidation of wastes. When properly operated, secondary treatment plants remove about 90 percent of the BOD and 90 percent of the suspended solids.

Advanced treatment processes are specialized, and their use depends on the pollutants that need to be removed. While usually advanced treatment follows primary and secondary treatment, in some cases (especially in industrial waste treatment), advanced treatment replaces conventional processes completely.

Collection Systems

Wastewater collection systems carry wastewater (along with the solids accumulated in it) from the source (residential, commercial, or industrial) to the treatment facility for processing. Modern, fully enclosed sewage systems ensure that water contaminated with wastes and pollutants does not pose heath, safety, or environmental problems.

To handle the needs of a service area (the area that a sewerage system will service), to take advantage of gravity and the natural drainage afforded by area geography, and to lessen the costs of installing lift stations or pumps to move waste flows, treatment facilities are usually constructed in or near low-lying community outskirts, frequently along the edge of a natural waterway (Parcher, 1998).

Collection System Types

Three types of sewerage systems are in general use: sanitary sewers, storm sewers, and combined sewer systems that carry both sanitary and stormwater flows.

1. **Sanitary sewers:** Sanitary sewers (which, by definition, carry human wastes) convey wastewater from residences, businesses, and some industries to the treatment facility. Unlike industrial waste flows, which may have some treatment prior to entry into a municipal system, the wastes that sanitary sewerage carries are untreated. Of primary concern in sanitary sewerage management is preventing sewerage overflows, since these wastes contain infectious materials that if released into the environment cause serious risks to public health.
2. **Storm sewers:** Storm sewers handle the influx of water into a collection system from surface runoff as the result of rainstorms or snowmelt. The more highly an area is developed, the more important effective stormwater collection systems are. As buildings and impermeable surfaces cover more area within a community, opportunities for storm flow to percolate into the ground to recharge groundwater are reduced, the surface runoff becomes heavier, and the number of contaminants and pollutants that can be carried by runoff are increased. Storm systems must be designed to handle sudden heavy flows that can contain large quantities of sand, silt,

grit, and gravel, as well as plant materials and trash. As long as these flows do not carry infectious or human wastes, storm sewers can often be shunted, untreated, to natural drainage, although primary treatment may be required to meet NPDES permit requirements.

3. **Combined sewerage systems:** Combined sewerage systems carry both sanitary and stormwater flows. Combination systems always carry sanitary wastes, but are designed to handle large flows as well (commonly up to three times the average flows), so that during heavy rainfall, the same sewerage system can handle stormwater runoff as well as the normal sanitary flows.

While some older systems are still in operation, combined sewers are now seldom installed in the United States because heavy precipitation can overwhelm the system, causing flows that exceed the treatment facility's ability to effectively treat them. Combined sewer overflows present a serious threat to public health.

Collection System Components

A community's sewerage system consists of the following:

- Building services that carry the wastes from the generation point to mains
- Mains that carry the wastes to collection sewers
- Collectors or subcollectors that carry the wastes to trunk lines
- Trunk lines that carry the wastewater flow to interceptors
- Interceptors that carry the wastes to the treatment plant
- Other systems elements that may include lift stations, manholes, vents, junction boxes, and cleanout points

Except for building connections, all of these components are built under streets, easements, and right-of-ways, on layouts that take into consideration ground elevation, gradient, and natural drainage. They are designed to meet considerations of population size, estimated flow rates, minimum and maximum loads, velocity, slope, depth, and the need for additional system elements to ensure adequate system flows and access for maintenance.

- **Lift Stations:** At points where gravity's force isn't enough to move wastewater through a system, lift stations are installed to pump the wastes to a higher point through a force main. Municipalities try to avoid installing lift stations; installation, operation, and maintenance of lift stations are expensive.
- **Manholes:** Access into the sewerage system for inspection, preventive maintenance, and repair is provided by manholes at regular intervals.

- **Vents:** Gases that build up within sewer systems from the wastes they carry must be vented safely from the system. Human wastes in sanitary sewers carry sulfides, and hydrogen sulfide is deadly. Industrial wastes can carry risks related to the composition of their wastewater components.
- **Junction boxes:** Sewerage systems are a network of piping, moving from small pipes that carry wastes from individual services to larger mains, collectors, trunk lines, and interceptors. The constructions that occur when individual lines join are junction boxes or chambers. They are of special concern because leakage and infiltration commonly occur at system joints.
- **Cleanout points:** Effective cleanout points provide access for cleaning equipment and maintenance into the sewer system.

Construction Materials

Sewer lines are made from a wide variety of materials, and construction material selection is based on a variety of possible conditions and factors.

- Rigid piping
- Cast iron, ductile iron, corrugated steel, sheet steel
- Concrete, reinforced concrete, asbestos cement
- Vitrified clay, brick masonry
- Flexible piping
- Plastics, including PVC, CPVC, and other thermoset plastics; polyfins, polyethelene, and other thermoplastics

Iron and steel piping offers the advantage of strength but is affected by corrosion both from the wastes the pipes carry and from soil conditions. These materials are frequently used in exterior spans (piping runs that bridge gullies, for example). Concrete types offer high strength for heavy loading, especially for large-volume pipes, but is heavy. Only short lengths of pipe are possible, so pipe runs must have many joints. (Joints provide areas in sewerage systems that allow for the advent of potential weak points through general deterioration, shifting, and root growth.) Clay is highly resistant to corrosion but is heavy and brittle. This type of piping is limited in length as well.

Plastics offer advantages of corrosion resistance, high strength-to-weight ratios, ease of handling, long pipe runs (so fewer joints), impermeability, and a certain amount of flexibility without loss of strength. They must be bedded carefully to avoid damage caused by soil voids.

Selection Factors

- Resistance of the material to corrosion
- Resistance of the material to flow and scour

- Resistance of the material to external and internal pressure
- Soil conditions and backfill
- The potential wastewater load's chemical make-up
- Requirements pertaining to strength, useful life, joint tightness, infiltration and inflow control, and other physical considerations
- Costs, availability, and ease of installation
- Ease of maintenance

New Technologies

New equipment and technology are rapidly changing the ways that collection systems are inspected, maintained, and repaired. Electronic tools have increased visual access to the sewer lines and have made possible increased use of more accurate mapping and database tools. Trenchless technologies are changing the methods by which sewerage systems are repaired or upgraded.

Electronic Technology for Collection Systems

The task of visually inspecting sewer lines presents some challenges. While sewer lamping, manhole inspection, and large sewer main entries are still used, technologies of miniaturization have provided collection systems with new tools to inspect and evaluate sewer lines, including visual entry into piping that, in the past, was too small to view effectively.

Specialized video equipment allows operators to assess line conditions quickly and accurately. With a waterproof video camera, lights, a system to transport the camera through the lines, and a closed-circuit TV system, sewer line conditions can be viewed, evaluated, and recorded. With sophisticated camera control equipment and robotics, operators can finally get a good look at problems far out of visual reach.

Video technology is not the only modern technology changing the way sewerage systems are managed. Computer record keeping, maintenance programs, and databases allow easy access to more information than was previously possible. Computer mapping techniques ensure that the information operators take out on the job is current and accurate. Geographic Information Systems provide spatial referencing that was not feasible only a few years ago.

Wastewater collection begins wastewater's journey from potable water use and discard to outfall. Once wastewater is collected, the processes necessary for treatment can begin.

REFERENCES

American Water Works Association. 1984. *Introduction to Water Treatment*, Vol. 2. Denver: American Water Works Association.

Lewis, S. A. 1996. *The Sierra Club Guide to Safe Drinking Water.* San Francisco: Sierra Club Books.

McGhee, T. J. 1991. *Water Supply and Sewerage.* 6th ed. New York: McGraw-Hill, Inc.

Parcher, M. J. 1998. *Wastewater Collection Systems Maintenance.* Lancaster, PA: Technomic Publishing Company, Inc.

Pielou, E. C. 1998. *Fresh Water.* Chicago: University of Chicago Press.

U.S. EPA. 2002. *Nonpoint Source News-Notes* 68 (June).

3

Homeland Security Strategy: Water/Wastewater

America is no longer protected by vast oceans. We are protected from attack only by vigorous action abroad, and increased vigilance at home.

—*President George W. Bush, January 29, 2002*

The terrorist events of 9/11 triggered immediate unease about distant concerns. In regard to the safety and security of our nation and its critical infrastructure, 9/11 precipitated the ultimate paradigm shift. For example, in President George W. Bush's statement above, he pointed out that "we are protected from attack only by vigorous action abroad;" thus, he orchestrated and directed a relentless attack on terrorists, other cold-blooded killers, and terrorism supporters residing in faraway lands including Afghanistan, Iraq, and selected locations throughout the globe.

The fact that we are presently fighting a global War on Terrorism is well known—the news media ensures that we are inundated (rightfully so) with this information on a daily basis. Televised coverage of our "Shock and Awe" campaigns in Afghanistan and Iraq are familiar to most Americans. Moreover, most Americans are aware of the government's ongoing campaign to hunt down and bring to justice the terrorists, wherever they reside.

While most Americans are well aware of our pursuit of the terrorists overseas, many have little knowledge of the government's efforts to protect the homeland. This is not to say that the average American has not heard about homeland security and some of its domestic antiterrorism policies—anyone who has boarded and flown in a commercial airplane since 9/11 is well aware of security changes at airports. What is not well known, however, is America's effort to protect its critical infrastructure, including its water/wastewater systems.

In June 2002, President Bush made an effort to publicize homeland security and the new Homeland Security Department, a cabinet-level position, by traveling to Kansas

City and focusing attention on the importance of placing domestic security under one head. During his Kansas City visit, President Bush visited a water treatment plant, which also focused attention on the importance of protecting the nation's critical water infrastructure.

Right after 9/11, many professional water/wastewater organizations were quick to put together and distribute guidance on protecting water and wastewater operations throughout the nation. Based on personal experience, and to my surprise, many of these directives were available to safety/security professionals in water/wastewater treatment operations within days after 9/11.

Within the following year (relatively quick for a major governmental agency), the Environmental Protection Agency (EPA) got on board and issued its Homeland Security Strategy, which was outlined in the document *Strategic Objectives in Homeland Security.* (The EPA is the lead governmental agency in protecting the nation's critical water/wastewater infrastructure.) In issuing this important document in September 2002, Christine Todd Whitman, EPA administrator, stated the following:

> The terrorist attacks of September 11, 2001, transformed the Environmental Protection Agency's long-standing mission to protect the environment and safeguard human health in new and important ways. For more than 30 years, the EPA has worked on behalf of the American people to protect our country from the effects of pollution and the threat of environmental degradation. Our goal has always been to make America's air cleaner, its water purer, and its land better protected. With the United States under threat of attack from international terrorists and others who seek to do our country harm, EPA's traditional mission has expanded to include protecting out country against the environmental and health consequences of acts of terrorism.

THE EPA'S STRATEGIC OBJECTIVES: HOMELAND SECURITY

In 2002, the Bush Administration developed a road map for securing the homeland—the National Strategy for Homeland Security (U.S. EPA, 2004)—which lays out specific objectives for border and transportation security, emergency preparedness and response, protection for critical infrastructure, domestic counterterrorism, defense against catastrophic threats, and intelligence and warning. This road map designates the EPA as the lead federal agency for protecting critical drinking water and wastewater treatment and distribution/collection infrastructure.

The EPA's work in water security is mandated by the Public Health Security and Bioterrorism Preparedness and Response Act (Bioterrorism Act) of 2002 (U.S. Congress, 2002). This law, coupled with executive directives and the agency's own strategic plan for homeland security, guide the agency's research and technical support activities to protect water infrastructure. The Homeland Security Presidential Directive on Critical Infrastructure Identification, Prioritization, and Protection (HSPD-7) reinforces the EPA's role as the sector-specific lead for water infrastructure (White House Office of Homeland Security, 2002). Moreover, it also assigns the responsibility of coordinating the overall national effort to protect critical infrastructure and key resources of the United States to the Department of Homeland Security.

Because of its sector-specific federal lead for protecting the nation's drinking water and wastewater infrastructures, The EPA plays a critical role in the homeland security arena—the EPA established the Water Protection Task Force to meet these responsibilities. In 2003, to provide research and technical support for the drinking water and wastewater sectors, the task force was organized formally as the Water Security Division and the National Homeland Security Research Center was established.

WATER SECURITY

In the following sections, water security directives and statutes (under Homeland Security legislation) are presented independent of wastewater security because the government's initial emphasis and rulemaking was targeted specifically at protecting the nation's drinking water supply and distribution system. Wastewater treatment and collection, on the other hand, was included as an afterthought primarily through the urging of several professional agencies in the wastewater field.

Public Health Security and Bioterrorism Preparedness and Response Act of 2002

On June 12, 2002, President Bush signed the Public Health Security and Bioterrorism Preparedness and Response Act of 2002 (Bioterrorism Act) into law (Public Law 107-188, 2002). The Bioterrorism Act amends the Safe Drinking Water Act (SDWA) by adding Section 1433. Section 1433(a) requires that certain community water systems (CWSs) conduct Vulnerability Assessments (VAs), certify to the EPA that the VAs were conducted, and submit a copy of the VA to the EPA. Section 1433(b) requires that certain CWSs prepare or revise Emergency Response Plans (ERPs) and certify to the EPA that an ERP has been completed.

Note that the EPA provided water/wastewater-specific guidance on how to comply with the Public Health and Security and Bioterrorism Preparedness and Response Act of 2002 with respect to the certification and submission of VAs and certification of completion of ERPs to the EPA.

Safe Drinking Water Act (SDWA): Section 1433

In the Bioterrorism Act, Congress recognizes the need for drinking water systems to undertake a more comprehensive view of safety and security—what this text refers to as super-sized S. The act amends the SDWA (Sections 1433[a] and [b]) and specifies actions community water systems and the EPA must take to improve the security of the nation's drinking water infrastructure.

The requirements of Sections 1433, 1434, and 1435 are listed in the appendix. It is important to note, however, that more comprehensive, in-depth discussions of VA and ERP requirements are covered in chapters 5 and 8, respectively.

WASTEWATER SECURITY

In March 2006, the U.S. Government Accountability Office (GAO) reported to the U.S. Senate that federal law does not address wastewater security as comprehensively as it does drinking water security. For example, wastewater facilities are not required by law to complete VAs. The Clean Air Act (CAA) does require wastewater facilities using certain amounts (above a threshold quantity) of hazardous substances, such as chlorine gas, sulfur dioxide gas, anhydrous and aqua ammonia, and so on, to submit to the EPA an RMP that details the accident prevention and emergency response activities. Also, under the EPA's guidance, the Clean Water State Revolving Fund (CWSRF) can be used in many instances for certain wastewater system security enhancements. While federal law governing wastewater security is limited, in December 2003, the president issued HSPD-7. The directive designated the EPA as the lead agency to oversee the security of the water sector, including both drinking water and wastewater critical infrastructures.

As mentioned earlier, Congress passed the Bioterrorism Act, which amended various laws, including the SDWA. The Bioterrorism Act required drinking water systems serving more than 3,300 people to complete VAs of their facilities by June 2004 and to prepare or update an existing ERP.

In 2003, Congress considered a bill that would have encouraged or required wastewater treatment plants to assess the vulnerability of wastewater facilities, make physical security improvements, and conduct research. However, the legislation did not become law and, consequently, no such requirement or specific funding exists for wastewater facilities.

Unfortunately, because Congress did not mandate specific security measures or provide the funding for security improvements at wastewater treatment facilities, many facilities have made security decisions based solely on local requirements, the prerogative of individual utility management personnel, and the availability of funding. Various wastewater industry professional organizations have partially filled the gap in lack of government guidance by providing written and computerized sample VAs and ERPs. At the present time, the shortfall that still exists in ensuring the security of wastewater facilities (relative to drinking water systems) is primarily from a lack of federal funding.

Notwithstanding the lack of federal law requiring wastewater systems to take security measures to protect specifically against a terrorist attack, as pointed out earlier, certain wastewater facilities are required to take security precautions employing the use of site-specific risk management plans (RMPs) that could mitigate the consequences of such an attack. Specifically, the EPA regulations implementing the CAA require these facilities to prepare RMPs that summarize the potential threat of sudden, accidental, large releases of certain chemicals, including the results that could occur off site in a worst-case chemical accident and the facility's plan to prevent releases and mitigate any damage. RMPs are to be revised and resubmitted to the EPA at least every five years, and the EPA is required to review them and require revisions, if necessary.

RMPs can be an effective tool in protecting wastewater facilities against terrorism. For example, the CAA imposes certain requirements on wastewater facilities regarding accidental releases. The act defines an *accidental release* as an unanticipated emission of a regulated substance or other extremely hazardous substance into the air, so any chemical release caused by a terrorist attack could be considered unanticipated and covered under the CAA. Such an interpretation would provide the EPA with authority under the act's RMP provisions to require security measures or vulnerability assessments with regard to terrorism. Moreover, Section 112(r)(1) of the CAA includes a general duty clause directing owners and operators of facilities that produce, process, handle, or store listed or other extremely hazardous substances to identify hazards, design and maintain a safe facility to prevent releases, and minimize the consequences of any accidental releases that occur. However, to date, the EPA has not attempted to use these CAA provisions because it is concerned that more specific legislation would pose significant litigation risk and has concluded that chemical facility security would be more effectively addressed by passage of specific legislation.

Based on personal experience, an RMP can be an effective vehicle for implementing security at wastewater facilities, but its success depends on the imagination of the implementer. However, it should be pointed out that compliance with RMP is expensive and requires almost daily vigilance to ensure full, ongoing compliance.

The Resource Conservation and Recovery Act (RCRA; also known as the Cradle to Grave Act) is another federal regulation affecting wastewater facilities that store certain amounts of hazardous chemicals. Though less effective in protecting facilities from intruders (terrorists included) than RMP, RCRA does require facilities that house hazardous waste to take certain security actions, such as posting warning signs, using a 24-hour surveillance system, or surrounding the active portion of the facility with a barrier and controlled entry gates.

While job security has always been an important factor in employee satisfaction, since 9/11 the term has taken on new meaning. Employees want to be reassured not only that their position with their employer is safe, but also that they are physically safe while on the job. Violence in the workplace, whether generated from inside or outside the workplace, is a growing problem. The possibility of terrorist attacks has only added to the burden of employers to protect their employees from harm.

In regards to preventing terrorist attacks on wastewater facilities, the likelihood of such attacks might be incidentally reduced by compliance with other federal statutes that impose safety requirements on certain wastewater facilities. These federal statutes also mitigate the consequences of terrorist attacks. For example, the Occupational Safety and Health (OSH) Act imposes a number of safety requirements, including a general duty to furnish a workplace free from recognized hazards that may cause death or serious physical harm to employees. This is known as the General Duty Clause. It is Section 5(a)1 of the OSH Act (OSHA, 2006), which states: "Each employer shall furnish to each of his [*sic*] employees employment and a place of employment which is free from recognized hazards that are causing or are likely to cause death or serious physical harm to his employees."

Another federal statute that can provide some security against terrorism or the impact of terrorism in wastewater facilities is the Emergency Planning and Community Right-to-Know Act (Public Law 99-499, 1986). The act requires owners of facilities that maintain specified quantities of certain extremely hazardous chemicals to submit information annually on their chemical inventory to state and local emergency response officials. The act also requires that each state establish a State Emergency Response Commission to oversee local emergency planning and create local emergency planning committees. These committees must develop and periodically review their communities' ERPs, including the identification of chemical facilities, and outline procedures for response personnel to follow in the event of a chemical incident.

Aside from statutes that address some areas of wastewater security, the EPA has asserted that federal funding is available for wastewater security-related measures though the CWSRF program (U.S. EPA, 2003). The EPA's CWSRF program provides grant funding to sites to allow them to assist publicly owned treatment works (POTWs) to make infrastructure improvements needed to protect public health and ensure compliance with the Clean Water Act. States may use CWSRF monies to provide low or zero percent interest rate loans to municipalities for wastewater infrastructure, including facility and sewer construction and rehabilitation, storm water management, and, combined, sewer and sanitary overflow correction.

This CWSRF funding information is important because anyone tasked with overseeing the upgrading of security in wastewater facilities quickly discovers that security improvements are all about the availability of lots of funding. Why? Because no one ever said security was cheap—it is not. CWSRF can provide the funding of security upgrades for wastewater facilities. For example, states may provide CWSRF assistance to POTWs to allow them to complete vulnerability assessments and contingency and emergency response plans. Many of the types of infrastructure improvements a wastewater facility might need to make to ensure security are also eligible for CWSRF funding and may have already been included within the scope of infrastructure projects

Table 3.1 Security Measures and Eligibility under the CWSRF Program

Type of Activity	*Eligible under CWSRF*
General	
Vulnerability assessments	Yes
Contingency/emergency response plans	Yes
Facility	
Security guards	No
Fencing	Yes
Security cameras/lighting	Yes
Motion detectors	Yes
Redundancy (systems and power)	Yes
Secure chemical and fuel storage	Yes
Lab equipment	Yes
Monitoring	No
Sewer System	
Securing large sanitary sewers	Yes
Tamper-proof manholes	Yes

Source: U.S. EPA, 2002.

funded through the program to date. Table 3.1 identifies specific activities that POTWs could take to ensure the security of their systems and indicates if the activity would be eligible through the CWSRF program.

Maintaining a human presence can be the most important security measure a POTW can take to ensure that its facilities are protected. However, as shown in table 3.1, the CWSRF program does not fund operations and maintenance activities for POTWs and therefore does not provide financing for an increased human security presence. CWSRF does, however, provide other infrastructure improvements that may be eligible for funding, such as the conversion from gaseous chemicals to alternative treatment processes, installation of fencing or security cameras, the securing of large sanitary sewers, and installment of tamper-proof manholes (U.S. EPA, 2002).

Another source of federal funding potentially available for wastewater security-related measures is the State Homeland Security Grant Program administered by the Department of Homeland Security. This program's primary objectives are to enhance the capacity of state and local emergency responders to prevent, protect against, respond to, and recover from terrorist incidents involving chemical, biological, radiological, nuclear, and explosive devices; agriculture; and cyber attacks. Under the program, grants are provided to states for a variety of purposes, including homeland security-related training and protection of critical infrastructure, although authority to make physical security improvements is limited. States are required to allocate at least 80 percent of these grant funds to local units of government, which include water districts (Government Accountability Office, 2006).

REFERENCES

OSH Act. 1970. U.S. Department of Labor, at www.osha.gov.

National Strategy for Homeland Security. 2002. White House Office of Homeland Security, at www.whitehouse.gov/homeland/book/nat_strat_his.pdf.

Public Law 99-499, 1986, 42nd Cong., §§ 11001-11050.

Public Health Security and Bioterrorism Preparedness and Response Act. 2002. Public Law 107-188, Cong., at www.epa.gov/safewater/watersecurity/pubs/security-act.pdf.

U. S. Environmental Protection Agency. 2002. *Clean Water Fact Sheet.* Washington, DC: Environmental Protection Agency.

U.S. Environmental Protection Agency. 2003. *Clean Water State Revolving Fund.* Washington, DC: Environmental Protection Agency, 832-F-03-002.

U.S. Environmental Protection Agency. 2004. *Water Security Research and Technical Support Action Plan.* Washington, DC: Environmental Protection Agency, 600/4-04/063.

U S. Government Accountability Office. 2006. *Securing Wastewater Facilities.* GAO-06-390.

4

Security Philosophy: Know Thine Enemy

> The emergence of amorphous and largely unknown terrorist individuals and groups operating independently (freelancers) and the new recruitment patterns of some groups, such as recruiting suicide commandos, female and child terrorists, and scientists capable of developing weapons of mass destruction, provide a measure of urgency to increasing our understanding of the psychological and sociological dynamics of terrorist groups and individuals.
>
> —*Rex A. Hudson, 1999*

If we (nonterrorists) were to qualitatively measure our understanding of a terrorist, the mental images that would most likely come to mind would be men, women, or children with assault rifles and bombs. We would most likely see men and women who feel so betrayed by something that they will attack anyone, anywhere, at any time, even the innocent, on their deathbeds, in a hospital, to get the recognition they think they deserve. We would most likely see bitter-hearted individuals, breeding bitterness and hatred.

If we were to quantitatively measure our understanding of why terrorists (or anyone else for that matter) desire to inflict death and destruction on those of us who live in freedom, we would score low on any scale. If we were to ask the average American, for example, why four groups of individuals banded together to hijack airplanes and then fly them, loaded with fuel and innocent passengers, into skyscrapers, the Pentagon, and attempt to strike other well-known targets, the standard answer might be: "They were insane."

Were the 9/11 terrorists insane? If we choose to answer this question in the affirmative, we are simply taking the easy way out, proving that we are ignoring the biblical injunction: "Know thine enemy." The fact is that it is easier for us to recognize terrorism than it is to define it. Similarly, it is almost impossible for any of us to identify a terrorist by sight alone. It is important for us to remember not only that biblical injunction but also the words of Dostoyevsky: "While nothing is easier than to denounce the evildoer, nothing is more difficult than to understand him." If we are to be successful in carrying out the War on Terrorism, in protecting our homeland, and in protecting our critical infrastructure, we must make every effort to understand the evildoers—the terrorists.

In our attempt to understand the terrorist, to know where he is coming from (and hopefully what he is up to), we would be well served by remembering some basic facts. For example, Erik Erikson (1994) points out that "Men who share an ethnic area, a historical era, or an economic pursuit are guided by common images of good and evil. Infinitely varied, these images reflect the elusive notion of historical change; yet in the form of contemporary social model, of compelling prototypes of good and evil, they assume decisive concreteness in every individual's ego development. Psychoanalysts' ego psychology has not matched this concreteness with sufficient theoretical specificity." In other words, there is more to the terrorist and to terrorism, in general, than meets the eye. We simply do not know what we do not know about the terrorist or terrorism. To be successful in our War on Terrorism, we must grasp the terrorist's mindset.

GETTING INTO THE MIND OF THE TERRORISTS

During an American, college-level class discussion on current affairs of safety and security issues in the workplace, the students were asked to describe terrorists and their mindset—to get into the mind of the terrorists. The students' responses are listed below (Old Dominion University, 2006).

Student 1: How can we get into the minds of madmen who decapitate people on television?

Student 2: People who kill innocent women and children to make some murky point do not have minds—they're simply crazy.

Student 3: Killing people in the name of religion, or some other warped belief, just does not make sense.

Student 4: I have no response, because I don't get it.

Student 5: Ditto that.

Student 6: My view is that they are jealous of the way we live and since they figure they can't live as we do, they will do everything possible to make us pay for it.

Student 7: We will never understand the terrorists' mindset because we do not understand why anyone would want to kill others and then themselves to make a point.

Student 8: I have no idea . . . They're just crazy, that's all.

Student 9: It's all about the Crusades, man. Like, when Western countries, at the direction of a delusional pope, sent hordes of criminals and anyone else they could gather up, to the Holy Land to free it from the Muslims . . . The fact of the matter is . . . that is, from what I know . . . the real reason the Crusades were put together was to capture Jerusalem and then dig into all those tunnels beneath the holy sites to find all that gold and treasure the Jews had hidden there years before.

Student 10: Well, I don't know anything about the Crusades and gold and treasure, but I do know that there is no way we will ever understand evil . . . We have to offset it with good . . . That's what the Bible says, I think.

Student 11: I do not want to get into the mind of terrorists . . . I do not want to think like them . . . I detest them and whatever it is they stand for.

Student 12: The terrorists are cowards . . . They do not fight fair . . . They kill anyone who happens to be around or gets in the way . . . They are natural born killers . . . That's all.

Students 13–18: These students had no comment.

After reviewing the unscientifically polled students' opinions above, which of the replies is closest to the point—closest to explaining what the terrorist's mindset is all about? Answer: Take your pick. We all have an opinion on the subject. Unfortunately, based on the students' replies and those of the so-called experts, we simply do not understand what we need to understand about terrorism and terrorists in general to properly defend ourselves against their actions. Even from this small pool of students, it is quite obvious that many of us find it easier to recognize terrorism but have difficulty in defining it. We simply try to sum up terrorism as insanity gone amok.

In the first place, terrorists are not insane, mad, or crazy. Instead, they are cold, calculating, and driven. We could say, rightfully so and based on a common viewpoint, that they are "misdriven" or "misguided," at best.

One of the responses that might garner your attention is the reply of student 12 who stated that "Terrorists are cowards . . . They do not fight fair." This seems to be the view of many Americans and others. However, consider the terrorists' plight, their point of view. We are the strongest nation on the globe. When we fight, we use the most sophisticated weapons available. We use highly trained soldiers, donned in state-of-the-art body armor. We use electronic warfare systems such as infrared detectors. We use the latest, greatest communication techniques available. Finally, in our minds we are on the side of right, and the terrorists are on the wrong side. There is no way a band of ill-equipped crazies can fight us man-to-man on any battlefield. Right? Absolutely, and herein lies the essence of the terrorists' strategy. That is, the terrorists' view is that "We (the terrorists) can't defeat coalition forces on their terms—on conventional terms. Thus, we choose to defeat them on our own terms."

The obvious question: What are the terrorists' terms?

If you accept the premise that terrorists are cold, calculating, and not crazy or insane, then it becomes a bit easier to see through their methodology, strategy, and/or tactics. For example, terrorists are cold because they justify the killing of innocents by professing that some innocents might indeed be innocent, but upon their death, they will have the distinct honor and pleasure of entering paradise, which (according to the terrorists) is a better place to reside than Earth; they are doing the innocents a big favor—just collateral damage with a huge benefit. In regard to the other innocents killed, terrorists will tell us that they are probably evil people and therefore Allah is being served by killing them. It is this mindset that provides the justification to the terrorist for killing anyone and everyone, including themselves—the rewards waiting for them in paradise are just too great.

To understand how terrorists are calculating, all one needs to do is view what is currently occurring in Iraq and Afghanistan. The most powerful military force on earth had little difficulty in liberating both Iraq and Afghanistan from standing forces that faced them head on. However, we continue to fight a nasty group of persistent insurgents (terrorists) on a daily basis.

Because the terrorists are calculating (actually quite innovative; they are definitely opportunists), they fully understand that if they band together and fight head-to-head against us (the coalition), they lose. Notwithstanding the antics of Richard Reid, the so-called shoe bomber, the terrorists are not stupid. Because they are not stupid, the terrorists that we are currently facing have taken up the tactics of guerrilla warfare. Using these tactics, they operate with small, mobile, and flexible combat groups called cells, without a front line.

In regard to tactics, terrorist actions are based on intelligence, ambush, deception, sabotage, and espionage, and their ultimate objective is usually to destabilize an authority through long, low-intensity confrontation. Does this sound familiar? Keep in mind that in the terrorist's view, two factors are critical to their success: (1) massive media coverage by a so-called bleeding-heart (anti-American) press displaying the gruesome terrorist acts to a worldwide audience, and (2) the ability to make it too expensive in manpower, treasury, and political terms for the occupier to occupy "their" land in a sustained manner. It is interesting to note, however, that guerrilla tactics against native regimes are generally unsuccessful.

The terrorists are innovative and opportunistic because they have an uncanny knack for devising very effective, offbeat killing devices from leftovers. Consider those items that are now familiar to many people: improvised explosive devices, or IEDs. IEDs are homemade booby traps manufactured to harass, maim, or kill. Typically thrown or laid on the sides of roads or hidden on the roads themselves, IEDs may be created using various household chemicals or military components in the right combustible combinations. Various packages can also be used to deliver the bomb from paper bags and steel pipes (pipe bombs) to cardboard boxes and milk containers. Other packages that have been used include mortar and howitzer shells. In addition, ingredients as simple as cooking oil have also been used in IEDs. These devices can be set off (detonated) by a timer, a timed fuse, or a cell phone. To date, in Iraq, the use of IEDs has been blamed for more deaths and casualties than any other terrorist tactic.

The bottom line: If we accept the premise that terrorists are cold, calculating innovators and opportunists, does this help us better understand terrorists and terrorism? Probably not. Simply, Americans have difficulty understanding terrorists or terrorism. Because of our failure to understand terrorists and terrorism, many experts state that to gain understanding, we must get into the mind of the terrorist. Get into the mind of the terrorist? How many of us can get into the minds of our spouses, children, coworkers, employers, or next-door neighbors? If you have the unique ability to do this, then you need go no further in this text. However, if you are like the rest of us and do not have a clue on how to get inside the mind of anyone, then you need to proceed on with this work, and hopefully you will at least have a basic understanding of how to protect water/wastewater infrastructure from a terrorist attack.

So, what is next? What is the road to understanding? Maybe we can get there if we equate terrorists and terrorism with evildoers and evil—Americans certainly understand these terms. Don't we? Maybe before we shift gears, it would be wise to regard the following:

Woe to those who call evil good, and good evil;
Who put darkness for light, and light for darkness;
Who put bitter for sweet, and sweet for bitter!

—*Isaiah 5:20*

Whenever we do not understand something, we tend to turn to the "experts" and their explanations for understanding. Let's get the social sciences point of view on terrorists and terrorism.

TERRORISM: THE SOCIAL SCIENCES VIEW

As mentioned, it is the view of this text that most terrorists, especially internationals, can be characterized as being cold, calculating innovators and opportunists. This characterization, however, does not explain the behavior or the actions of terrorists. But does any characterization explain their behavior or actions? Usually, when we have questions about human behavior, we defer to the psychologists. But can psychology, the study of the mind, tell us what motivates terrorists? More importantly, can psychology tell us how the terrorists differ from the rest of us?

In attempting to answer these questions, we have a problem: The terrorists don't normally volunteer for psychological analyses. As you might imagine, psychologists and other terrorism experts differ in their opinions on terrorism and terrorists in general. The psychologist Rona Fields (1979), for example, has psychologically tested terrorists from Northern Ireland, Lebanon, Southeast Asia, Israel, and Africa. She thinks the Khmer Rouge, who created a bloodbath in Cambodia during the 1970s, are similar in mindset to today's suicide bombers—they share the same stillborn moral and emotional development. Fields notes, "Their definition of right and wrong is very black-and-white, and is directed by an authoritative director. . . . There's a total limitation of the capacity to think for themselves. [Terrorists] believe there's a difference between right and wrong, but when they do something in the name of the cause, it's justified."

Richard Pearlstein (1991), associate professor of political science at Southeastern Oklahoma State University, states that terrorists "are rational, they are not insane. They have goals and they are moving towards those goals."

David Long (1990), former assistant director of the State Department's Office of Counter Terrorism, writes that not only are terrorists not crazy, but they also don't share a personality type. He says, "No comparative work on terrorist psychology has ever succeeded in revealing a particular psychological type or uniform terrorist type."

Another terrorism expert, Rex A. Hudson, a researcher and author for the Library of Congress, published a prestigious report entitled, *The Sociology and Psychology of Terrorism: Who Becomes a Terrorist and Why?* This interesting and eye-opening report should be mandatory reading for anyone involved with ensuring the safety/security of any facility. In this September 1999 report, Hudson's predictions of future terrorist acts against America are uncanny in their accuracy concerning the foretelling of a 9/11-type event. For example, among other projections, Hudson speculated that an al Qaeda-like cell might load passenger airplanes with high explosives and deliberately fly them into the Pentagon, White House, and other important buildings. In his report, Hudson did not mention the Twin Towers of the World Trade Center as a possible target. However, he certainly presaged many of the other terrorist actions.

So what are Hudson's views on terrorism? In his report, Hudson points out that "definitions of terrorism vary widely and are usually inadequate." He goes on to point out that it is much easier to define a terrorist action. According to Hudson, "A terrorist action is the calculated use of unexpected, shocking, and unlawful violence against noncombatants (including, in addition to civilians, off-duty military and security personnel in peaceful situations) and other symbolic targets perpetrated by a clandestine member(s) of a subnational group or a clandestine agent(s) for the psychological purpose of publicizing a political or religious cause and/or intimidating or coercing a government(s) or civilian population into accepting demands on behalf of the cause."

Hudson's definition of a terrorist is somewhat similar to the definition used in this text (i.e., we describe terrorists as being cold, calculating innovators and opportunists). According to Hudson, "[Terrorists] are people with cunning, skill, and initiative, as well as ruthlessness."

Hypotheses of Terrorism

Earlier it was stated, in general, that terrorists are not insane. If one accepts this proposition, then one has to ask him or herself the following questions: What causes anyone to become a terrorist? Is it nature or nurture? Are they made or born? This is a difficult question to definitively answer simply because full-scale, quantitative studies from which to develop specific answers on terrorism are lacking, and almost nonexistent. However, this does not preclude the development of several theories on the topic. One theory, the Mancur Olson (1971) hypothesis, suggests that participants in revolutionary violence predicate their behavior on a rational, cost-benefit calculation and the conclusion that violence is the best available course of action given the social conditions. This text does not buy into this theory simply because it is questionable that any group rationally chooses a terrorism strategy. Hudson (1999) points out that "a group's decision to resort to terrorism is often divisive, sometimes resulting in fractionalization of the group."

Frustration-Aggression (F-A) Hypothesis

The frustration-aggression (F-A) hypothesis of violence was first developed by a group of social psychologists in 1939 and later proposed by Ted Robert Gurr (1970). F-A states that frustration causes aggression and that catharsis is the reduction in the aggressive drive following an aggressive act. Frustration (interference with goal-directed behavior) arouses a drive whose primary goal is that of harming a person or object—usually the perceived cause of the frustration. Joseph Margolin (1977), a proponent of this hypothesis, argues that "much terrorist behavior is a response to the frustration of various political, economic, and personal needs or objectives."

Hudson (1999) points out that a better approach than these and other hypotheses would be the subcultural theory presented by Franco Ferracuti (1982), which takes into account that terrorists live in their own subculture with their own value systems. Similarly, based on observation and common sense, frustration does not always lead to aggression, nor is aggression always the result of frustration. Frustration only leads to aggression when it is seen as intentionally meant to thwart the person. Further, Paul

Wilkinson (1974) faults the F-A hypothesis for having "very little to say about the social psychology of prejudice and hatred" and fanaticisms that "play a major role in encouraging extreme violence." Wilkinson believes that "Political terrorism cannot be understood outside the context of the development of terrorist, or potentially terroristic, ideologies, beliefs and lifestyles."

Negative Identity Hypothesis

The term *negative identity* is derived from Erikson's (1994) theory of identity formation. Basically, an individual with a negative identity engages in negative behavior. For example, a terrorist, negative in identity, engages in negative behaviors such as hijacking airplanes or bombing schoolhouses full of children and teachers. Negative identity involves a vindictive rejection of the role regarded as desirable and proper by the individual's family and community (Hudson, 1999).

Narcissistic Rage Hypothesis

The narcissistic rage hypothesis is all about terrorists being mentally ill, as a result of events that occurred in early childhood. Those possessing this self-love and narcissus trait react to anyone who threatens their inflated self-image. Secretly, they have doubts about the accuracy of this self-image. John Crayton (1983) states that "as a specific manifestation of narcissistic rage, terrorism occurs in the context of narcissistic injury." For Crayton, terrorism is an attempt to acquire or maintain power or control by intimidation (Hudson, 1999).

THE 411 ON SAFETY AND SECURITY

If we accept the psychological axiom, "The only problem that cannot be solved is the problem of refusing solutions," as the final statement relating to combating terrorism, then our ongoing fight against terrorists and terrorism was lost before it began. The problem with terrorists is that they absolutely demand nothing less than for someone else to change. Even when threatened with death, the terrorist will refuse negotiation, because negotiation would be the ultimate personal humiliation.

Most of this chapter to this point has dealt with the social scientists' views on terrorists and terrorism. Most so-called experts in the field of trying to define terrorists and terrorism state that we can't do that unless we get into the mind of the terrorist. In this text, it has been made clear that getting into the mind of anyone is beyond human capability. So is all the social science literature and research that attempts to define or to get into the mind of the terrorist just nonsense? That opinion or judgment is left to the reader to make.

If it is true, as this text professes, that we can't get into the mind of the terrorist, then what is the answer? How do we protect ourselves? How do we counteract the terrorist and, ultimately, terrorism?

The answer is that we can only succeed in our War on Terrorism if we physically and mentally switch places with the terrorist. That's right; we must put ourselves in the terrorist's shoes. This scenario is particularly effective when the goal is to protect critical infrastructure, such as water/wastewater systems, from terrorist attack.

How does switching places work? Good question. The approach the author used in 2001 (and subsequently in auditing various other water/wastewater infrastructure), one week after 9/11, was to visit each wastewater treatment plant and other piece of critical wastewater infrastructure and just stand outside the facility, looking in, from several locations along the periphery. I asked myself: If I was a terrorist (or any other cold-blooded criminal, for that matter), how would I attack this site, facility, pumping station, underground sewer, chemical storage area, railroad chemical tank car, and so on? Having put on hold all aspects of personal decency and compassion, I simply turned a cold and numb mind to observing the obvious—breaking in and doing serious damage at these sites absolutely was no problem. The realization of what I found was actually more chilling than my evil thoughts. A terrorist would have had a field day blowing up any of the unit processes and structures and systems that I was responsible for—just a simple cake walk in the deliverance of death and destruction.

I have used this same switching procedure for the last few years at various other water/wastewater facilities, and others—including water storage tanks and reservoirs—and have found it effective in guarding against the terrorist or other intruder to the best level possible.

Is it foolproof? No; nothing is foolproof. However, when you switch places with the terrorist or intruder and actually get outside the box (not just think outside the box) and actually look from the outside to the inside, you can observe what needs to be observed. This observation effort is greatly enhanced if you record all that you observe. For this reason, it is a good idea to have not only another set of eyes (a safety or security professional) with you, but also a person to record what is observed.

Some would say that the procedure described above is nothing more than the performance of a vulnerability assessment (VA). This is not true. The VA, using a checklist, is the second step in this procedure. The first step is to step into the shoes of the terrorist looking inward; and the second step is to bounce these observations against a well-formulated VA. Finally, after observation and the VA, follow-up (or action—the most difficult part) is required.

In the chapter to follow, more is said about VAs. For the present, it is important to stress not only the importance of not thinking inside the box, but also of physically getting outside the box and looking inward. Hopefully, it is only from the outside looking in that we are concerned about. However, reality being what it is, we cannot overlook the possibility of terrorism from within, including violence in the workplace. This is the worst-case scenario, obviously. However, it is a possibility that cannot be overlooked or disregarded. In this age of cyberspace and disgruntled employees, the cyber jockey (or other employee) you work with every day could be a terrorist in waiting—ready to spring into action. Thus, we must not only get outside the box, but also get inside the box and look thoroughly in all directions.

REFERENCES

Crayton, J. W. 1983. Terrorism and the psychology of the self. In *Perspectives on Terrorism*, ed. L. Z. Freedman and Y. Alexander, 33–41. Wilmington, DE: Scholarly Resources.

Erikson, E. 1994. *Identity of the Life Cycle.* New York: W.W. Norton & Company.

Ferracuti, F. 1982. A sociopsychiatric interpretation of terrorism. *The Annals of the American Academy of Political and Social Science* 463 (September): 129–41.

Fields, R. M. 1979. Child terror victims and adult terrorist. *Journal of Psychohistory* 7(1).

Gurr, T. R. 1971. *Why Men Rebel.* Princeton, NJ: Princeton University Press.

Hudson, R. A. 1999. *The Sociology and Psychology of Terrorism: Who Becomes a Terrorist and Why?* Washington, DC: Library of Congress.

Long, D. E. 1990. *The Anatomy of Terrorism.* New York: Free Press.

Margolin, J. 1977. Psychological perspectives on terrorism. In *Terrorism: Interdisciplinary Perspectives*, ed. Y. Alexander and S. M. Finger. New York: John Jay Press.

Olson, M. 1971. *The Logic of Collective Action.* Boston: Harvard University Press.

Pearlstein, R. 1991. *The Mind of the Political Terrorist.* Wilmington, DE: Scholarly Resources.

Spellman, F. R. 2006. Safety/security principles. Lecture presented at Old Dominion University, Norfolk, VA.

Wilkinson, P. 1974. *Political Terrorism.* London: Macmillan.

5

Vulnerability Assessment

Vulnerability means different things to different people. Some view it as an intrinsic characteristic of soils and other parts of the natural environment. Others find that vulnerability depends on the properties of individual contaminants or contaminant groups, but is independent of specific land-use or management practices. Still others associate vulnerability with a specific set of human activities.

—*Stephen Foster, 1987*

One consequence of the events of September 11, 2001, was the Environmental Protection Agency's (EPA) directive to establish a Water Protection Task Force to ensure that activities to protect and secure water supply/wastewater treatment infrastructure are comprehensive and carried out expeditiously. Another consequence is a heightened concern among citizens in the United States over the security of their critical water/wastewater infrastructure. As mentioned, the nation's water/wastewater infrastructure—consisting of several thousand publicly owned water/wastewater treatment plants, more than 100,000 pumping stations, hundreds of thousands of miles of water distribution and sanitary sewers, and another 200,000 miles of storm sewers—is one of America's most valuable resources, with treatment and distribution/collection systems valued at more than $2.5 trillion. Wastewater treatment operations taken alone include the sanitary and storm sewers forming an extensive network that runs near or beneath key buildings and roads and is contiguous to many communication and transportation networks. Significant damage to the nation's wastewater facilities or collection systems would result in loss of life; catastrophic environmental damage to rivers, lakes, and wetlands; contamination of drinking water supplies; long term public health impacts; destruction of fish and shellfish production; and disruption to commerce, the economy, and our normal way of life.

VULNERABILITY ASSESSMENT (VA): WATER/WASTEWATER

The water sector was designated by President Clinton's (1998) Presidential Decision Directive 63 and reaffirmed by President Bush's (2001) Executive Order 13231 as one of eight critical infrastructures deemed essential to the nation's well-being. In response to the Public Health Security and Bioterrorism Preparedness and Response Act of 2002, the EPA has developed baseline threat information to use in conjunction with vulnerability assessments. Furthermore, to defray some of the cost of those studies, the EPA has provided assistance to drinking water systems to enable them to undertake vulnerability assessments and develop emergency response plans (see chapter 6).

Post-9/11, the water sector has taken great strides to protect its critical infrastructure. For instance, government and industry have developed vulnerability assessment methodologies for both drinking water and wastewater facilities and trained thousands of utility operators to conduct them. Again, however, it is important to point out that those lawmakers and regulatory agencies initially were concerned primarily with drinking water safety/security. The importance of including wastewater operations in water sector security provisions—similar to what was required for drinking water infrastructure—was quickly recognized by several professional associations involved in wastewater system operations. These organizations made every effort to voice their concerns about the need to include wastewater infrastructure in the EPA's requirement to perform a vulnerability assessment (VA). To date, the EPA has mentioned the importance of including wastewater systems within the blanket provision of water sector security, but the VAs officially published by the regulators still remain tailored for drinking water systems. Therefore, many utility systems, of their own volition and primarily at their own expense, have incorporated wastewater infrastructure into VAs performed within their water sector security responsibilities.

Because of the obvious gap in specific regulations that leaves wastewater systems to their own devices for ensuring their own security, the EPA and industry professionals have ensured those responsible for ensuring safety/security of wastewater infrastructure that many of the necessary safety/security steps are covered by the EPA's Risk Management Program (RMP). Having conducted VAs and modified VAs (wastewater) on both water and wastewater systems and having implemented RMP in a major U.S. wastewater system, I have to concur with this point of view. Note, however, that RMP applies to those wastewater facilities that store, use, or produce (e.g., digester gas) covered hazardous materials (chlorine, ammonia, sulfur dioxide, etc.) at or above a designated threshold quantity (TQ) level. Also note that the trend in recent years has been to replace or substitute these covered hazardous materials with nonhazardous materials (e.g., replacing covered chlorine with uncovered sodium hypochlorite).

Points of Vulnerability: Water/Wastewater Systems

Many of the water/wastewater system components that combine to make up critical infrastructure, such as pumping stations, maintenance centers, wastewater interceptor/water distribution lines, and tanks/reservoirs, are located in isolated locations. Water/wastewater distribution and interceptor networks, therefore, have an inherent potential to be vulnerable to the full spectrum of threats: physical, biological, chemical, and nuclear. Any of these threats, obviously, could compromise the system's ability to reliably deliver safe water and/or protect water supplies and the environment. These areas of vulnerability include (1) the raw water source (surface or groundwater), (2) raw water channels and pipelines, (3) raw water reservoirs, (4) water/wastewater treatment facilities, (5) connections to the distribution system, (6) pump stations and valves, (7) sewers, (8) interceptor lines, and (7) finished water tanks and reservoirs. Each of these system components and unit processes presents a unique challenge to the water utility in safeguarding the entire water supply (Clark and Deininger, 2000).

What Is Vulnerability Assessment?

The complexity of vulnerability assessments will vary based on the design and operation of the water/wastewater system. The nature and extent of the VA will differ among systems based on a number of factors, including system size, potential population, and safety. Evaluations also vary based on knowledge and types of threats; available security technologies; and applicable local, state, and federal regulations. Preferably, a VA is performance based, meaning that it evaluates the risk to the water/wastewater system based on the effectiveness (performance) of existing and planned measures to counteract adversarial actions. According to Larry Mays and the U.S. EPA (2004; 2002), the common elements of water/modified wastewater vulnerability assessments include the following (wastewater added):

- Characterization of the water/wastewater system, including its mission and objectives
- Identification and prioritization of adverse consequences to avoid
- Determination of critical assets that might be subject to malevolent acts that could result in undesired consequences
- Assessment of the likelihood (qualitative probability) of such malevolent acts from adversaries
- Evaluation of existing countermeasures
- Analysis of current risk and development of a prioritized plan for risk reduction

Vulnerability Assessment Tools

Several vulnerability assessment tools are available. Two of the most commonly used tools are the Risk Assessment Methodology for Water Utilities and Vulnerability Self-Assessment Tool.

Water ISAC: Information Sharing

The Water ISAC is a highly secure, Internet-based communications tool, developed to provide America's drinking and wastewater systems with a secure web-based environment for early warning of potential threats and a source of knowledge about water system security. The Water ISAC is open to all U.S. drinking water and wastewater systems. The Water ISAC serves three purposes:

- To disseminate early warnings and alerts concerning threats against the physical infrastructure and cyber systems of drinking water and wastewater utilities
- To allow drinking water and wastewater utilities to share with each other information on security incidents
- To provide an opportunity for utilities to have security incidents analyzed by counterterrorism experts

The types of information provided include the following:

- Potential and imminent threats to utilities
- Reports of incidents nationwide
- Incident trends
- Possible responses to threats and attacks
- Research on infrastructure protection

Information sources include the following:

- Water utilities
- Counterterrorism experts
- Federal law enforcement agencies
- Federal intelligence agencies
- The EPA
- Industry research

Table 5.1 presents a sampling of website locations provided by Water ISAC

Points to Consider in a Vulnerability Assessment

Some points to consider related to the six basic elements listed above are included in table 5.2. The manner in which the vulnerability assessment is performed is determined by each individual water/wastewater utility. Throughout the assessment process, it is important to remember that the ultimate goal is threefold: to safeguard public health and safety, to reduce the potential for disruption of a reliable supply of pressurized water, and to ensure proper disposal of wastewater (U.S. EPA, 2006).

Table 5.1. Website Locations Provided by Water ISAC

Water Sector Links	
American Water Works Association	http://www.awwa.org
Association of State Drinking Water Administrators	http://www.asdwa.org
Water Environment Federation	http://www.wef.org
Contaminant Resources	
Centers for Disease Control Biological Agents	http://www.bt.cdc.gov/agent
Federal Links	
Department of Homeland Security	http://www.dhs.ogv
Environmental Protection Agency	http://www.epa.gov/safewater
Federal Bureau of Investigation (FBI)	http://www.fbi.gov
Other Links	
International Association of Emergency Managers	http://www.iaem.com
National Drinking Water Clearinghouse	http://www.ndwc.wvu.edu

Table 5.2. Basic Elements in Vulnerability Assessments

Element	Points to Consider
1. Characterization of the water/wastewater system, including its mission and objective	What are the important missions of the system to be assessed? Define the highest priority services provided by the utility. Identify the utility's customers: • General public • Government • Military • Industrial • Critical care • Retail operations • Firefighting What are the most important facilities, processes, and assets of the system for achieving the mission objectives and avoiding undesired consequences? Describe the following: • Utility facilities • Operating procedures • Management practices that are necessary to achieve the mission objectives • How the utility operates (e.g., water source including ground and surface water) • Treatment processes • Storage methods and capacity • Chemical use and storage • Distribution system In assessing those assets that are critical, consider critical customers, dependence on other infrastructures (e.g., electricity, transportation, other water utilities), contractual obligations, single points of failure (e.g., critical aqueducts, transmission systems, aquifers, etc.), chemical hazards and other aspects of the

Table 5.2. (*Continued*)

Element	Points to Consider
	utility's operations, or availability of other utility capabilities that may increase or decrease the criticality of specific facilities, processes, and assets.
2. Identification and prioritization of adverse consequences to avoid	Take into account the impacts that could substantially disrupt the ability of the system to provide a safe and reliable supply of drinking water and disposal of wastewater or otherwise present significant public health concerns to the surrounding community. Water/wastewater systems should use the vulnerability assessment process to determine how to reduce risks associated with the consequences of significant concern. Ranges of consequences or impacts for each of these events should be identified and defined. Factors to be considered in assessing the consequences may include the following: • Magnitude of service disruption • Economic impact (such as replacement and installation costs for damaged critical assets or loss of revenue due to service outage) • Number of illnesses or deaths resulting from an event • Impact on public confidence in the water supply • Chronic problems arising from specific events • Other indicators of the impact of each event as determined by the water/wastewater utility Risk reduction recommendations at the conclusion of the vulnerability assessment strive to prevent or reduce each of these consequences.
3. Determination of critical assets that might be subject to malevolent acts	What are the malevolent acts that could reasonably cause undesired consequences? Consider the operation of critical facilities, assets, and/or processes and assess what an adversary could do to disrupt these operations. Such acts may include physical damage to or destruction of critical assets, contamination of water, intentional release of stored chemicals, interruption of electricity, or other infrastructure interdependencies. The Public Health Security and Bioterrorism Preparedness and Response Act of 2002 (PL 107-188) states that a community water system which serves a population of greater than 3,300 people must review the vulnerability of its system to a terrorist attack or other intentional acts intended to substantially disrupt the ability of the system to provide a safe and reliable

supply of drinking water. The vulnerability assessment shall include, but not be limited to, a review of the following:

- Pipes and constructed conveyances
- Physical barriers
- Water/wastewater collection, pretreatment, and treatment facilities
- Storage and distribution facilities
- Electronic, computer, or other automated systems that are utilized by the public water system (e.g., Supervisory Control and Data Acquisition [SCADA])
- The use, storage, or handling of various chemicals
- The operation and maintenance of such systems

4. Assessment of the likelihood of malevolent acts from terrorists and vandals

Determine the possible modes of attack that might result in consequences of significant concern based on critical assets of the water system. The objective of this step of the assessment is to move beyond what is merely possible and determine the likelihood of a particular attack scenario. This is a very difficult task, as there is often insufficient information to determine the likelihood of a particular event with any degree of certainty.

The threats (the kind of adversary and the mode of attack) selected for consideration during a vulnerability assessment will dictate, to a great extent, the risk reduction measures that should be designed to counter the threat(s). Some vulnerability assessment methodologies refer to this as a Design Basis Threat (DBT), where the threat serves as the basis for the design of countermeasures, as well as the benchmark against which vulnerabilities are assessed. It should be noted that there is no single DBT or threat profile for all water systems in the United States. Differences in geographic location, size of the utility, previous attacks in the local area, and many other factors will influence the threat(s) that water/wastewater systems should consider in their assessments. Water/wastewater systems should consult with the local FBI and/or other law enforcement agencies, public officials, and others to determine the threats upon which their risk reduction measures should be based. Water/wastewater systems should also refer to the EPA's "Baseline Threat Information for Vulnerability Assessments of Community Water Systems" to help assess the most likely threats to their system. This document is available to community water systems serving populations greater than 3,300 people. If your system has not yet received instructions on how to receive a

Table 5.2. (***Continued***)

Element	Points to Consider
	copy of this document, then contact your regional EPA office immediately. You will be sent instructions on how to securely access the document via the Water Information Sharing and Analysis Center (ISAC) website or obtain a hardcopy that can be mailed directly to you. Water systems may also want to review their incident reports to better understand past breaches of security. Note: Water ISAC provides a secure forum for gathering, analyzing, and sharing security-related information. Additionally, several federal agencies are working together to improve the warehousing of information regarding contamination threats, such as the release of biological, chemical, and radiological substance into the water supply, and ways to respond to their presence in drinking water. With respect to identifying new technologies, the EPA has an existing program that develops testing protocols and verifies the performance of innovative technologies. It has also initiated a new program to verify monitoring technologies that may be useful in detecting or avoiding biological or chemical threats.
5. Evaluation of existing countermeasures	What capabilities does the system currently employ for detection, delay, and response? • Identify and evaluate current detection capabilities such as intrusion detection systems, water quality monitoring, operational alarms, guard post orders, and employee security awareness programs. • Identify current delay mechanisms such as locks and key control, fencing, structure integrity of critical assets, and vehicle access checkpoints. • Identify existing policies and procedures for evaluation and response to intrusion and system malfunction alarms, adverse water quality indicators, and cyber system intrusions. • It is important to determine the performance characteristics. Poorly operated and maintained security technologies provide little or no protection. What cyber protection system features does the utility have in place? Assess what protective measures are in place for the SCADA and business-related computer information systems such as firewalls, modem access, Internet, and other external connections, including wireless data and voice communications and security policies and protocols. It is important to identify

whether vendors have access rights and/or "backdoors" to conduct system diagnostics remotely.

What security policies and procedures exist, and what is the compliance record for them? Identify existing policies and procedures concerning the following:

- Personal security
- Physical security
- Key and access badge control
- Control of system configuration and operational data
- Chemical and other vendor deliveries
- Security training and exercise records

6. Analysis of current risk and development of a prioritized plan for risk reduction

Information gathered on threats, critical assets, water utility operations, consequences, and existing countermeasures should be analyzed to determine the current level of risk. The utility should then determine whether current risks are acceptable or risk reduction measures should be pursued.

Recommended actions should measurably reduce risks by reducing vulnerabilities and/or consequences through improved deterrence, delay, detection, and/or response capabilities or by improving operational policies or procedures.

Selection of specific risk reduction actions should be completed prior to considering the cost of the recommended action(s). Utilities should carefully consider both short- and long-term solutions. An analysis of the cost of short- and long-term risk reduction actions may impact which actions the utility chooses to achieve its security goals.

Utilities may also want to consider security improvements. Security and general infrastructure may provide significant multiple benefits. For example, improved treatment processes or system redundancies can both reduce vulnerabilities and enhance day-to-day operation.

Generally, strategies for reducing vulnerabilities fall into three broad categories:

- Sound business practices—affect policies, procedures, and training to improve the overall security-related culture at the drinking water facility. For example, it is important to ensure rapid communication capabilities exist between public health authorities and local law enforcement and emergency responders.

Table 5.2. (*Continued*)

Element	Points to Consider
	• System upgrades—include changes in operations, equipment, processes, or infrastructure itself that make the system fundamentally safer. • Security upgrades—improve capabilities for detection, delay, or response.

Security Vulnerability Self-Assessment Guide for Small Drinking Water Systems

The EPA, in collaboration with the Association of State Drinking Water Administrators and National Rural Water Association (2002), produced a Security Vulnerability Self-Assessment Guide for Small Drinking Water Systems Serving Populations between 3,300 and 10,000 guidance document. This Security Vulnerability Self-Assessment Guide is designed to help small water systems determine possible vulnerable components and identify security measures that should be considered to protect the system and the customers it serves. As mentioned, a VA is the identification of weaknesses in water system security; a VA focuses on defined threats that could compromise water system's ability to meet its various service missions—such as providing adequate drinking water, water for firefighting, and/or water for various commercial and industrial purposes. This document is designed particularly for systems that serve populations of 3,300 to 10,000. This document is meant to encourage smaller systems to review their system vulnerabilities, but it may not take the place of a comprehensive review by security experts. Completion of this document will meet the requirement for conducting a VA as directed under the Public Health Security and Bioterrorism Preparedness and Response Act of 2002. Community Water Systems (CWSs) serving more than 3,300 and fewer than 50,000 people were required to submit their completed vulnerability assessments to the administrator of the EPA no later than June 30, 2004, in order to meet the provisions of the act.

As shown in table 5.3, this Self-Assessment Guide has a simple design. Answers to assessment questions are yes or no, and there is space to identify needed actions and actions that have been taken to improve security. For any no answer, refer to the comment column and/or contact your state drinking water primary agency.

In addition to the general checklist for your entire water system (questions 1–15), you should give special attention to the following issues, presented in sections of table 5.3, related to various water system components.

- **Water sources:** Your water sources (surface water intakes or wells) should be secured. Surface water supplies present the greatest challenge. Typically they encompass large land areas. Where areas cannot be secured, steps should be taken to initiate or increase law enforcement patrols. Pay particular attention to surface water intakes. Ask the public to be vigilant and report suspicious activity.

- **Treatment plant and suppliers:** Some small systems provide easy access to their water system for suppliers of equipment, chemicals, and other materials for the convenience of both parties. This practice should be discontinued.
- **Distribution:** Hydrants are highly visible and convenient entry points into the distribution system. Maintaining and monitoring positive pressure in your system is important to provide fire protection and prevent introduction of contaminants.
- **Personnel:** You should add security procedures to your personnel policies.
- **Information/storage/computers/controls/maps:** Security of the system, including computerized controls like a Supervisory Control and Data Acquisition system, goes beyond the physical aspects of operation. It also includes records and critical information that could be used by someone planning to disrupt or contaminate your water system.
- **Public relations:** You should educate your customers about your system. You should encourage them to be alert and to report any suspicious activity to law enforcement authorities.

Table 5.3. Security Vulnerability Self-assessment for Small Water Systems

The first 15 questions in this vulnerability self-assessment are general questions designed to apply to all components of your system (wellhead or surface water intake, treatment plant, storage tank(s), pumps, distribution system, and offices). These are followed by more specific questions that look at individual system components in greater detail.

Question	*Answer*	*Comment – Action Taken/Needed*
	GENERAL	
1. Do you have a written emergency response plan (ERP)?	Yes/No	Under the provisions of the Public Health Security and Bioterrorism Preparedness and Response Act of 2002, you are required to develop and/or update an ERP within six months after completing this assessment. If you do not have an ERP, you can obtain a sample from your state drinking water primacy agency. As a first step in developing your ERP, you should develop your emergency contact list. A plan is vital in case there is an incident that requires immediate response. Your plan should be reviewed at least annually (or more frequently if necessary) to ensure it is up-to-date and addresses security emergencies including ready access to laboratories capable of analyzing water samples. You should coordinate with your LEPC.

Table 5.3. (*Continued*)

Question	*Answer*	*Comment – Action Taken/Needed*
		You should designate someone to be contacted in case of emergency regardless of the day of the week or time of day. This contact information should be kept up-to-date and made available to all water system personnel and local officials (if applicable). Share this ERP with police, emergency personnel, and your state primacy agency. Posting contact information is a good idea only if authorized personnel are the only ones seeing the information. These signs could pose a security risk if posted for public viewing since they give people information that could be used against the system.
2. Have you reviewed the EPA's baseline threat information document?	Yes/No	The EPA baseline threat document is available through the Water Information Sharing and Analysis Center at www.waterisac.org. It is important you use this document to determine potential threats to your system and to obtain additional security-related information. The EPA should have provided a certified letter to your system that provided instructions on obtaining the threat document.
3. Is access to the critical components of the water system (i.e., the parts of the physical infrastructure that are essential for water flow and quality) restricted to authorized personnel only?	Yes/No	You should restrict or limit access to the critical components of your water system to authorized personnel only. This is the first step in security enhancement for your water system. Consider the following: • Issue water system photo identification cards for employees and require them to be displayed within the restricted area at all times. • Post signs restricting entry to authorized personnel and ensure that assigned staff escort people without a proper ID.
4. Are all critical facilities fenced, including well houses and pump pits, and are gates locked where appropriate?	Yes/No	Ideally, all facilities should have a security fence around the perimeter. The fence perimeter should be walked periodically to check for breaches and maintenance needs. All gates should be locked with chains and a tamper-proof padlock that, at a minimum, protects the shank. Other barriers, such as concrete jersey barriers, should be considered to guard certain critical components from accidental or intentional vehicle intrusion.
5. Are critical doors, windows, and other points of entry, such as tank and roof hatches and vents, kept closed and locked?	Yes/No	Lock all building doors and windows, hatches and vents, gates, and other points of entry to prevent access by unauthorized personnel. Check locks regularly. Deadbolt locks and lock guards provide a high level of security for the cost.

A daily check of critical system components enhances security and ensures that an unauthorized entry has not taken place.

Doors and hinges to critical facilities should be constructed of heavy-duty, reinforced material. Hinges on all outside doors should be located on the inside.

To limit access to water systems, all windows should be locked and reinforced with wire mesh or iron bars and bolted on the inside. Systems should ensure that this type of security meets the requirements of any fire codes. Alarms can also be installed on windows, doors, and other points of entry.

6. Is there external lighting around all critical components of your water system? Yes/No

Adequate lighting of the exterior of the critical components is a good deterrent to unauthorized access and may result in the detection of such access. Motion detectors that activate lights or trigger alarms also enhance security.

7. Are warning sings (No Tampering, Unauthorized Access, etc.) posted on all critical components, such as well and storage tanks? Yes/No

Warning signs are an effective means to deter unauthorized access. A sign stating, "Warning: Tampering with this facility is a federal offense" should be posted on all water facilities. These are available from your state rural water association. Other signs may include: "Authorized Personnel Only," "Unauthorized Access Prohibited," and "Employees Only."

8. Do you patrol and inspect all source intake, buildings, storage tanks, equipment, and other critical components? Yes/No

Frequent and random patrolling of the water system by utility staff may discourage potential tampering. It may also help identify problems that may have arisen since the previous patrol.

All systems are encouraged to initiate personal contact with the local law enforcement to show them the drinking water facility. The tour should include the identification of all critical components with an explanation of why they are important. Systems are encouraged to review, with local law enforcement, the NRWA/ASDWA Guide for Security Decisions or similar state document to clarify respective roles and responsibilities in the event of an incident. Also consider asking the local law enforcement to conduct periodic patrols of your water system.

9. Is the area around critical components free of objects that may be used for breaking and entering? Yes/No

When assessing the area around critical components, look for large rocks, cement blocks, pieces of wood, ladders, valve keys, and other tools.

10. Are the entry points to your system easily seen? Yes/No

You should clear fence lines of all vegetation. Overhanging or nearby trees also provide easy access. Avoid landscaping that will permit trespassers to hide or conduct unnoticed

Table5.3. (Continued)

Question	*Answer*	*Comment – Action Taken/Needed*
		suspicious activities. Trim trees and shrubs to enhance the visibility of your water system's critical components. If possible, park vehicles and equipment in places where they do not block the view of your water system's critical components.
11. Do you have an alarm system that will detect unauthorized or attempted entry?	Yes/No	Consider installing an alarm system that notifies the proper authorities or your water system's designated contact for emergencies when there has been a breach of security. Inexpensive systems are available. An alarm system should be considered whenever possible for tanks, pump houses, and treatment facilities. You should also have an audible alarm at the site as a deterrent and to notify neighbors of a potential threat.
12. Do you have a key control and accountability policy?	Yes/No	Keep a record of locks, associated keys, and to whom the keys have been assigned. This record will facilitate lock replacement and key management (e.g., after employee turnover or loss of keys). Vehicle and building keys should be kept in a lockbox when not in use. You should have all keys stamped (engraved) "DO NOT DUPLICATE."
13. Are entry codes and keys limited to water system personnel only?	Yes/No	Suppliers and personnel from colocated organizations (e.g., organizations using your facility for telecommunications) should be denied access to codes and/or keys. Codes should be changed frequently if possible. Entry into any building should always be under the direct control of water system personnel.
14. Do you have an updated operations and maintenance manual that includes evaluations of security systems?	Yes/No	Operation and maintenance plans are critical in assuring the ongoing provision of safe and reliable water service. These plans should be updated to incorporate security considerations and the ongoing reliability of security provisions—including security procedures and security-related equipment.
15. Do you have a neighborhood watch program for your water system?	Yes/No	Watchful neighbors can be very helpful. Make sure they know who to call in the event of an emergency activity.
	WATER SOURCES	
16. Are your wellheads sealed properly?	Yes/No	A properly sealed wellhead decreases the opportunity for the introduction of contaminants. If you are not sure whether your wellhead is properly sealed, contact your well drilling/maintenance company, your state drinking water primacy agency, your state rural water association, or other technical assistance providers.
17. Are well vents and caps screened and securely attached?	Yes/No	Properly installed vents and caps can help prevent the introduction of a contaminant into the water supply. Ensure that vents and caps

		serve their purpose and cannot be easily breached or removed.
18. Are observation, test, and abandoned wells properly secured to prevent tampering?	Yes/No	All observation/test and abandoned wells should be properly capped or secured to prevent the introduction of contaminants into the aquifer or water supply. Abandoned wells should be either removed or filled with concrete.
19. Is your surface water source secured with fences or gates? Do water system personnel visit the source?	Yes/No	Surface water supplies present the greatest challenge to security. Often, they encompass large land areas. Where areas cannot be secured, increase patrols by water utility personnel and law enforcement agents.
	TREATMENT PLANT AND SUPPLIERS	
20. Are deliveries of chemicals and other supplies made in the presence of water system personnel?	Yes/No	Establish a policy that an authorized person designated by the water system must accompany all deliveries. Verify the credentials of all drivers. This prevents unauthorized personnel from having access to the water system.
21. Have you discussed with your suppliers procedures to ensure the security of their products?	Yes/No	Verify that your suppliers take precautions to ensure that their products are not contaminated. Chain of custody procedures for delivery of chemicals should be reviewed. You should inspect chemicals and other supplies at the time of delivery to verify they are sealed and in unopened containers. Match all delivered goods with purchase orders to ensure that they were, in fact, ordered by your water system. Keep a log or journal of deliveries. It should include the drivers (taken from the driver's photo ID), date, time, material delivered, and the supplier's name.
22. Are chemicals, particularly those that are potentially hazardous (e.g., chlorine gas) or flammable, properly stored in a secure area?	Yes/No	All chemicals should be stored in an area designated for their storage only, and the area should be secure and access to the area restricted. Access to chemical storage should be available only to authorized employees. Pay special attention to the storage, handling, and security of chlorine gas because of its potential hazard. You should have tools and equipment on-site (such as a fire extinguisher, drysweep, etc.) to take immediate actions when responding to an emergency.
23. Do you monitor raw and treated water so that you can detect changes in water quality?	Yes/No	Monitoring of raw and treated water can establish a baseline that may allow you to know if there has been a contamination incident. Some parameters for raw water include pH, turbidity, total and fecal coliform, total organic carbon, specific conductivity, ultraviolet adsorption, color, and odor. Routine parameters for finished water and distribution systems include free and total chlorine residual, heterotrophic plate count

Table 5.3. (*Continued*)

Question	*Answer*	*Comment – Action Taken/Needed*
		(HPC), total and fecal coliform, pH, specific conductivity, color, taste, odor, and system pressure. Chlorine demand patterns can help you identify potential problems with your water. A sudden change in demand may be a good indicator of contamination in your system. For those systems that use chlorine, absence of chlorine residual may indicate possible contamination. Chorine residuals provide protection against bacterial and viral contamination that may enter the water supply.
24. Are tank ladders, access hatches, and entry points secured?	Yes/No	The use of tamper-proof padlocks at entry points (hatches, vents, and ladder enclosures) will reduce the potential for unauthorized entry. If you have towers, consider putting physical barriers on the legs to prevent unauthorized climbing.
25. Are vents and overflow pipes properly protected with screens or grates?	Yes/No	Air vents and overflow pipes are direct conduits to the finished water in storage facilities. Secure all vents and overflow pipes with heavy-duty screens and/or grates.
26. Can you isolate the storage tank from the rest of the system?	Yes/No	A water system should be able to take its storage tank(s) off-line if there is a contamination problem or structural damage. Install shutoff or bypass valves to allow you to isolate the storage tank in the case of a contamination problem or structural damage. Consider installing a sampling tap on the storage tank outlet to test water in the tank for possible contamination.
	DISTRIBUTION	
27. Do you regulate the use of hydrants and valves?	Yes/No	Your water system should have a policy that regulates the authorized use of hydrants for purposes other than fire protection. Require authorization and backflow devices if a hydrant is used for any purpose other than fire fighting. Consider designating specific hydrants for use as filling station(s) with proper backflow prevention (e.g., to meet the needs of construction firms). Then, notify local law enforcement officials and the public that these are the only sites designated for this use. Flush hydrants should be kept locked to prevent contaminants from being introduced into the distribution system and to prevent improper use.
28. Does your system monitor for and maintain positive pressure?	Yes/No	Positive pressure is essential for firefighting and for preventing back-siphonage that may

		contaminate finished water in the distribution system. Refer to your state primacy agency for minimum drinking water pressure requirements.
29. Has your system implemented a backflow prevention program?	Yes/No	In addition to maintaining positive pressure, backflow prevention programs provide an added margin of safety by helping to prevent the intentional introduction of contaminants. If you need information on backflow prevention programs, contact your state drinking water primacy agency.
	PERSONNEL	
30. When hiring personnel, do you request that local police perform a criminal background check, and do you verify employment eligibility as required by the INS Form 1-9?	Yes/No	It is good practice to have all job candidates fill out an employment application. You should verify professional references. Background checks conducted during the hiring process may prevent potential employee-related security issues. If you use contract personnel, check on the personnel practices of all providers to ensure that their hiring practices are consistent with good security practices.
31. Are your personnel issued photo ID cards?	Yes/No	For positive identification, all personnel should be issued water system photo ID cards and be required to wear them at all times. Photo identification will also facilitate identification of authorized water system personnel in the event of an emergency.
32. When terminating employment, do you require employees to turn in photo IDs, keys, access codes, and other security-related items?	Yes/No	Former or disgruntled employees have knowledge about the operation of your water system, and could have both the intent and physical capability to harm your system. Requiring employees who will no longer be working at your water system to turn in their IDs, keys, and access codes helps limit these types of security breaches.
33. Do you use uniforms and vehicles with your water system name prominently displayed?	Yes/No	Requiring personnel to wear uniforms and requiring that all vehicles prominently display the water system name helps inform the public when water system staff is working on the system. Any observed activity by personnel without uniforms should be regarded as suspicious. The public should be encouraged to report suspicious activity to law enforcement authorities.
34. Have water system personnel been advised to report security vulnerability concerns and to report suspicious activity?	Yes/No	Your personnel should be trained about issues at your facility, what to look for, and how to report any suspicious events or activity. Periodic meetings of authorized personnel should be held to discuss security issues.
35. Do your personnel have a checklist to use for threats or suspicious calls or to report suspicious activity?	Yes/No	To properly document suspicious or threatening phone calls or reports of suspicious activity, a simple checklist can be used to record and report all pertinent information. Calls should be reported immediately to appropriate law enforcement officials. Checklists should be available at every telephone. Also consider installing caller

Table 5.3. (*Continued*)

Question	*Answer*	*Comment – Action Taken/Needed*
		ID on your telephone system to keep a record of incoming calls.
	INFORMATION/STORAGE/COMPUTERS/CONTROLS/MAPS	
36. Is computer access password protected? Is virus protection installed and software upgraded regularly, and are your virus definitions updated at least daily? Do you have Internet firewall software? Do you have a plan to back up your computers?	Yes/No	All computer access should be password protected. Passwords should be changed every 90 days and (as needed) following employee turnover. When possible, each individual should have a unique password that they do not share with others. If you have Internet access, a firewall protection program should be installed on your side of the computer and reviewed and updated periodically. Also consider contracting a virus protection company and subscribing to a virus update program to protect your records. Backing up computers regularly will help prevent the loss of data in the event that your computer is damaged or breaks. Backup copies of computer data should be made routinely and stored at a secure, offsite location.
37. Is there information on the web that can be used to disrupt your system or contaminate your water?	Yes/No	Posting detailed information about your water system on a website may make the system more vulnerable to attack. Websites should be examined to determine whether they contain critical information that should be removed. You should do a web search (using a search engine such as Google, Yahoo!, or Lycos) using keywords related to your water supply to find any published data on the web that is easily accessible by someone who may want to damage your water supply.
38. Are maps, records, and other information stored in a secure location?	Yes/No	Records, maps, and other information should be stored in a secure location when not in use. Access should be limited to authorized personnel only. You should make back-up copies of all data and sensitive documents. These should be stored in a secure, offsite location on a regular basis.
39. Are copies of records, maps, and other sensitive information labeled confidential, and are all copies controlled and returned to the water system?	Yes/No	Sensitive documents (e.g., schematics, maps, and plans and specifications) distributed for construction projects or other uses should be recorded and recovered after use. You should discuss measures to safeguard your documents with bidders for new projects.
40. Are vehicles locked and secured at all times?	Yes/No	Vehicles are essential to any water system. They typically contain maps and other

		information about the operation of the water system. Water system personnel should exercise caution to ensure that this information is secure. Water system vehicles should be locked when they are not in use or left unattended. Remove any critical information about the system before parking vehicles for the night. Vehicles also usually contain tools (e.g., valve wrenches) and keys that could be used to access critical components of your water system. These should be secured and accounted for daily.
	PUBLIC RELATIONS	
41. Do you have a program to educate and encourage the public to be vigilant and report suspicious activity to assist in the security protection of your water systems?	Yes/No	Advise your customers and the public that your system has increased preventive security measures to protect the water supply from vandalism. Ask for their help. Provide customers with your telephone number and the telephone number of the local law enforcement authorities so that they can report suspicious activities. The telephone number can be made available through direct mail, billing inserts, notices on community bulletin boards, flyers, and consumer confidence reports.
42. Does your water system have a procedure to deal with public information requests and to restrict distribution of sensitive information?	Yes/No	You should have a procedure for personnel to follow when you receive an inquiry about the water system or its operation from the press, customers, or the general public. Your personnel should be advised not to speak to the media on behalf of the water system. Only one person should be designated as the spokesperson for the water system. Only that person should respond to media inquiries. You should establish a process for responding to inquiries from your customers and the general public.
43. Do you have a procedure in place to receive notification of a disease immediately after discovery by local health agencies?	Yes/No	It is critical to be able to receive information about suspected problems with the water at any time and respond to them quickly. Written procedures should be developed in advance with your state drinking water primacy agency, local health agencies, and your local emergency planning committee and reviewed periodically.
44. Do you have a procedure in place to advise the community of contamination immediately after discovery?	Yes/No	As soon as possible after a disease outbreak, you should notify testing personnel and your laboratory of the incident. In outbreaks caused by microbial contaminants, it is critical to discover the type of contaminant and its method of transport (water, food, etc.). Active testing of your water supply will enable your laboratory, working in conjunction with public

Table 5.3. *(Continued)*

Question	*Answer*	*Comment – Action Taken/Needed*
		health officials, to determine if there are any unique (and possibly lethal) disease organisms in your water supply. It is critical to be able to get the word out to your customers as soon as possible after discovering a health hazard in your water supply. In addition to your responsibility to protect public health, you must also comply with the requirements of the Public Notification Rule. Some simple methods include announcements via radio or television, door-to-door notification, a phone tree, and posting notices in public places. The announcement should include accepted uses for the water and advice on where to obtain safe drinking water. Call large facilities that have large populations of people who might be particularly threatened by the outbreak, such as hospitals, nursing homes, the school district, jails, large public buildings, and large companies. Enlist the support of local emergency response personnel to assist in the effort.
45. Do you have a procedure in place to respond immediately to customer complaints about a new taste, odor, color, or other physical change (oily, filmy, burns on contact with skin) in the water?	Yes/No	It is critical to be able to respond to and quickly identify potential water quality problems reported by customers. Procedures should be developed in advance to investigate and identify the cause of the problem, as well as to alert local health agencies, your state drinking water primacy agency, and your local emergency planning committee if you discover a problem.

RISK MANAGEMENT PROGRAM: WASTEWATER

Background

On November 10, 2005, Senator Jim Jeffords of Vermont introduced legislation that would authorize almost $300 million to bolster the safety and security of the nation's wastewater treatment plants. Jeffords said, "We know the potential dangers posed by the use of toxic chemicals at our nation's wastewater treatment plants, and our homeland security strategy should reflect those dangers. This bill takes the essential first step in closing the security gaps that make our wastewater treatment systems vulnerable to a terrorist attack" (2005).

In February 2005, the Government Accountability Office (GAO) released a report on wastewater security that ranks the release of chlorine as the second-highest security risk after damage to sewer collection systems. There have recently been two major accidents involving trains that were transporting chlorine, one of which resulted in nine deaths in South Carolina in January 2005.

The Wastewater Treatment Works Security Act of 2005 requires all wastewater facilities in the United States to conduct vulnerability assessments, develop site security and emergency response plans, and consider alternative approaches to potentially high-risk treatment methods. In addition, the bill provides millions of dollars in needed funds to prepare and implement plans, to research innovative technologies, and to assist small communities in complying with the requirements.

The Wastewater Treatment Works Security Act will codify what are now voluntary prevention and security measures and require all wastewater facilities to complete vulnerability assessments and emergency response plans, just as drinking water facilities have done since 2002.

Some of the particulars of the Jeffords act mentioned above might make one wonder why we are just now getting around to worrying about the security of wastewater treatment systems. What took so long? Since the events of 9/11, the security of the nation's drinking water and wastewater infrastructure has received increased attention from Congress and the executive branch. This is what is in print and what is said by government officials. But is this really the case?

After 9/11, when government officials were scrambling to identify critical infrastructures in the United States that needed to be protected and to identify which of these needed safety/security upgrades, the nation's water systems were listed in the top eight on every such list. Unfortunately, focus on wastewater systems took a backseat to drinking water systems. Why? Based on personal experience (and opinion), there was an initial mindset that terrorists wouldn't want to blow up sewage treatment plants. Many so-called experts in the field of safety/security scoffed at the idea of protecting such plants. This was the case even though many water/wastewater facilities were sited side-by-side. It makes little sense to cover the water side of the operation with codified security requirements and ignore the wastewater side. Wastewater treatment plants and associated infrastructure were like the bastard children of the water service industry, left to fend for themselves.

Fortunately, this convoluted mindset is changing—slowly. For example, in January 2005, the GAO published a study entitled *Wastewater Facilities: Experts' Views on How Federal Funds Should Be Spent to Improve Security*. The GAO found that experts identified the collection system's network of sewer lines as the most vulnerable asset of a wastewater utility. Experts state that the sewers could be used as a means to covertly gain access to surrounding buildings or as a conduit to inject hazardous substances that could impair a wastewater treatment plant's capabilities. Among the other vulnerabilities most frequently cited were the storage and transportation of chemicals used in the wastewater treatment process and the automated systems that control many vital operations. In addition, experts described a number of vulnerabilities that are not specific to particular assets but that may also affect the security of wastewater facilities. These vulnerabilities include a general lack of security awareness among wastewater facility staff and administrators, interdependencies among various wastewater facility components that make it possible for the disruption of a single component to devastate the entire system, and interdependencies between wastewater facilities and other critical infrastructure.

Experts identified several key activities as most deserving of federal funds to improve wastewater facilities' security. Among those most cited was the replacement of gaseous chemicals used in the disinfection process with less hazardous alternatives. This activity was rated as the highest priority for federal funding by 29 of 50 experts. Other security-enhancing activities most often rated as high priority included improving local, state, and regional collaboration (23 of 50 experts) and supporting facilities' efforts to comprehensively assess their vulnerabilities (20 of 50 experts).

Reality Check

Again, even with pending legislation such as Jeffords's act, homeland security provisions for protecting wastewater infrastructure have taken a backseat to drinking water security. This is not to say that all utility managers have failed to act on protecting wastewater infrastructure; those with the drive, initiative, and resources have acted. However, based on personal observation and experience, much of the wastewater infrastructure in the United States is still lacking in proper security measures.

Other than the direction and guidance provided by various wastewater professional organizations, the only regulatory requirements dealing with security for specified wastewater infrastructure (i.e., chemical storage and handling facilities) are provided by OSHA through its Process Safety Management Standard (PSM) (29 CFR 1910.119) and EPA's RMP (Part 68, Subpart G, Appendix F).

As pointed out earlier, OSHA's PSM was the first major step toward protecting wastewater facilities that use covered (listed) hazardous materials in their unit processes. The EPA borrowed the tenets of PSM and incorporated many of them into its RMP program. In the following section, the EPA's Supplemental Risk Management Program Guidance for Wastewater Treatment Plants is discussed.

Risk Management Program

Community residents and industry officials do not consider the importance of accident prevention unit after an accident occurs. . . . But then, the ghosts of Bhopal's victims must whisper, the only response can be. Too late. Too late.

—*Minter, 1996*

On May 24, 1996, the EPA finalized the RMP under Section 112(r) of the 1990 Clean Air Act Amendments. On June 20, 1996, the EPA promulgated the new rule. The rule, under 40 CFR Part 68, is entitled, "Accidental Release Prevention Provisions: Risk Management Programs." Covered sources have until June 1999 to complete data, devise a risk management plan, institute a risk management program to comply with RMP, and submit the risk management plan to the EPA for review and approval (Spellman, 1997).

Note: It should be pointed out that, like with OSHA's PSM, it is important to distinguish between having a plan and a program. Specifically, the *plan* is the information and the document that the facility submits to the regulatory agency (the EPA for RMP) and maintains onsite for use by facility personnel. The *program*, however, is the system that backs up the plan and helps to ensure that the facility is operated according to the rule. The viable program is more than just a vehicle to be used in improving the facility's safety profile; it should also provide dividends as regards productivity, efficiency, and profitability. It is important to keep in mind that to be beneficial (i.e., to reduce accidents and injuries), the program, as with any other management tool, must be upgraded and improved on a continuing basis.

RMP addresses specific chemicals/materials (compounds); it addresses the accidental release of over 100 chemical substances. Of the RMP chemicals listed, 77 are acutely toxic chemical compounds, 63 are flammable gases, and the rest are other federally listed chemical constituents. Threshold quantity levels range from 500 pounds to 20,000 pounds. The EPA estimates that approximately 100,000+ sources are covered by the rule. The sources include chemical and most other manufacturers, certain wholesalers and retailers, drinking water systems, water treatment works, ammonia refrigeration systems, chemical wholesalers and end users, utilities, propane retailers, and federal facilities.

General Applicability

Table 5.4 includes the chemicals regulated under Part 68 and OSHA PSM that are commonly used at wastewater treatment plants (WWTPs), along with their associated TQ.

For methane, the 10,000-pound threshold applies to the total weight of the flammable mixture of digester gases, not just the weight of methane or flammables in the mixture. However, if your WWTP uses methane (or a methane mixture) as fuel or sells it as fuel (as a retail facility), the amount that you use or sell as fuel is not covered under Part 68. For aqueous ammonia, the threshold applies only to the weight of ammonia in the mixture.

In general, regulated chemicals in a wastestream at a POTW will not exceed one percent of a mixture and thus will not be covered by Part 68. If an industrial WWTP has

Table 5.4. Threshold Quantities of Chemicals Used at Wastewater Treatment Plants

Chemical	*EPA Threshold Quantity*	*OSHA Threshold Quantity*
Chlorine	2,500 pounds	1,500 pounds
Anhydrous ammonia	10,000 pounds	10,000 pounds
Aqueous ammonia	20,000 pounds	15,000 pounds (>44%, concentration 20% or less)
Anhydrous sulfur dioxide	5,000 pounds	1,000 pounds (Liquid)
Methane	10,000 pounds	10,000 pounds

more than one percent of a regulated substance in the wastestream, the quantity of the substance in the wastestream will have to be determined and compared to the TQ. Oxygen is not subject to either Part 68 or OSHA PSM. Ozone is also not subject to Part 68, although it is covered by OSHA PSM.

RMP includes three major parts or elements. These important elements are hazard assessment, prevention program, and response program; each of these elements is addressed below.

Hazard Assessment A hazard assessment is required to assess the potential effects of an accidental (or intentional) release of a covered chemical/material. This RMP element generally includes performing an off-site consequence analysis (OCA) and the compilation of a five-year accident history. The OCA must include analysis of at least one worst-case scenario. This alternative scenario must be one that is more likely to occur than the worst-case scenario and that reaches an endpoint offsite; also, each covered toxic substance must have an alternative release scenario. The EPA has summarized some simplified consequence modeling approaches in an OCA guidance document. This OCA guidance document contains tables of dispersion and explosion modeling results that allow those who use them to minimize modeling efforts. In its modeling requirement, the EPA has specified numerous mandatory modeling parameters and assumptions, primarily for the worst-case scenario analyses, to make OCAs more consistent. The worst-case scenario release quantity is defined as the largest vessel or pipe inventory, taking into consideration administrative controls that could limit the maximum inventory before the release. Generally, gas releases are assumed to occur over a ten-minute period; liquid pools are assumed to form instantaneously and then vaporize. Passive mitigation system credit may be given, if the system is capable of withstanding the release event of interest. For flammable releases, the analyst must assume that the entire release quantity vaporizes and undergoes a vapor cloud explosion.

- **Worst-Case Scenario:** When considering the stationary source's worst-case scenario, there are selection factors to be considered. In addition to the largest inventories of a substance, the following conditions must also be considered: smaller quantities handled at higher process temperatures and pressures, and proximity to the boundary of the stationary source. Sources must analyze and report additional worst-case scenarios for a hazard class if the worst-case scenario from another covered process affects a different set of public receptors than the original worst-case scenario.

- **Alternative Release Scenario:** Alternative release scenarios must be more likely to occur than the worst-case scenario and must reach an endpoint offsite. In addition to those factors listed above, the EPA says owners should consider five-year accident history and failure scenarios identified by a process hazard analysis (PHA) or Program Level 2 hazard review when selecting alternative release scenarios. The alternative release scenario analyses may be performed using somewhat more flexible modeling approaches and parameters than those specified for worst-case scenario analyses. For example, active mitigation credit can be given.
- **Estimating Distances:** For both the worst-case and alternative release scenarios, the source must estimate the distance to where the endpoint is no longer exceeded and estimate the population (rounded to two significant digits) within a circle defined by the distance and centered at the release point. U.S. Census data may be used and they do not have to be updated; however, the presence of sensitive populations (hospitals, schools, etc.) must be noted. In addition, the source must identify and list the types of environmental receptors within the calculated worst-case distance and circle; however, no environmental damage assessment is required. In determining the presence of environmental receptors, U.S. Geological Survey maps may be used.

The off-site consequence analysis must be reviewed and updated every five years. However, if process changes might reasonably be expected to cause the worst-case scenario footprint or signature to increase or decrease by a factor of two or more, then the OCA must be revised, and the risk management plan must be resubmitted to the EPA or designated authority within six months.

It is important to note that the five-year history must cover all accidental releases from covered processes that resulted in deaths, injuries, or significant property damage onsite, as well as known off-site deaths, injuries, evacuations, sheltering in place, property damage, or environmental damage. The EPA requires that ten specific types of accident data be compiled, including known initiating event, off-site impacts, contribution factors, and operation or process changes that resulted from investigation of the release.

Prevention Program A prevention program is required to prevent accidental releases of regulated substances. This element generally includes safety precautions and maintenance, monitoring, employee safety training, and other requirements similar to OSHA's PSM. It should be pointed out, however, that the EPA's requirements for the

Program Level 2 elements (listed below) are less detailed than their OSHA PSM counterpart.

- Safety information
- Hazard reviews
- Compliance audits
- Maintenance
- Operating procedures
- Incident investigation
- Training

For example, the hazard review requirements vary from OSHA's process hazard analysis provision in the following ways:

- No team requirement for the review
- Fewer technical issues addressed in the analysis
- Results to be documented and problem resolved in a timely manner but no requirement for a formal resolution system
- No requirement to keep all hazard review results for the life of the process
- No requirement to communicate findings to employees

Although the prevention program language of RMP's Program Level 2 is somewhat different from the requirements in the OSHA PSM Standard, this is not the case with the language of RMP's Program Level; it is virtually identical to that of the OSHA PSM Standard, except that the RMP rule uses different terms for some things (to be discussed later)—these differences are based on the different legislative authorities that each agency holds. The EPA has also deleted specific phrases from the OSHA PSM regulatory language for the process safety information, process hazard analysis, and incident investigation elements to ensure that all sources implement process safety management in a way that protects not only workers, but also the public and the environment. Because of this language difference, companies should incorporate considerations of off-site effects into their OSHA PHA revalidation protocols.

Response Program The response program requires specific action to be taken in emergency situations. This element generally includes procedures for notifying public and local agencies responsible for responding to accidental releases, disseminating information on emergency health care, and training employees in emergency response. These employee response training measures are required for plants where employees are intended to respond to accidental releases using the plant's plan. The plan must address public notification, emergency medical treatment for accidental human exposures, and procedures of emergency response.

RMP Definitions and Requirements

The final management planning regulations (40 CFR Part 68) define the activities that sources must undertake to address the risks posed by regulated substances in covered processes. To ensure that individual processes are subject to appropriate requirements that match their size and the risks they may pose, the EPA has classified these activities into three categories, or programs. These program classifications are described below along with the requirements for regulated processes in each category.

Program 1

Requirements apply to processes for which a worst-case release, as evaluated in the hazard assessment, would not affect the public. These are sources or processes that have not had an accidental release that caused serious off-site consequences. Remotely located sources and processes using listed flammables are primarily those eligible for this program.

Program 1 Requirements

- **Hazard assessment:** worst-case analysis, five-year accident history
- **Prevention program:** certify no additional steps needed
- **Emergency response program:** coordinate with local responders
- **Risk management plan contents:** executive summary, registration, worst-case data, five-year accident history, certification

Program 2

Requirements apply to less complex operations that do not involve chemical processing (e.g., retailers, propane users, nonchemical manufacturers' processes not regulated under OSHA's PSM standard).

Program 2 Requirements

- **Hazard assessment:** worst-case analysis, alternative releases, five-year accident history
- **Management program:** document management system
- **Prevention program:** safety information, hazard review, operating procedures, training, maintenance, incident investigation, compliance audit
- **Emergency response program:** develop plan and program
- **Risk management plan contents:** executive summary, registration; worst-case data, alternative release data, five-year accident history, prevention program data, emergency response data, certification

Program 3

Requirements apply to higher-risk, complex chemical processing operations, to sources having a relevant process in one of nine named SIC codes listed in table 5.5, or to sources that have process(es) subject to the OSHA PSM.

Table 5.5. Program 3: SIC Code Applicability

SIC Code	*Industry*
2611	Pulp mills
2812	Alkalis and chlorine
2819	Industrial inorganics
2821	Plastics and resins
2865	Cyclic crudes
2869	Industrial organics
2873	Nitrogen fertilizers
2879	Agricultural chemicals
2911	Petroleum refineries

Program 3 Requirements

- **Hazard assessment:** worst-case analysis, alternative releases, five-year accident history
- **Management program:** document management system
- **Prevention program:** program safety information, process hazard analysis, operating procedures, training, mechanical integrity, incident investigation, compliance audit, management of change, prestartup review, contractors, employee participation, hot work permits
- **Emergency response program:** develop plan and program
- **Risk management plan contents:** executive summary, registration, worst-case data, alternative release data, five-year accident history, prevention program data, emergency response data, certification

RMP/PSM: Similarities (Overlap)

It would be incorrect to relate the comparison of RMP and PSM in the context of RMP versus PSM. The fact is that the RMP rule and PSM Standard are designed to work together; they complement each other. This can be seen quite clearly when the similarities of the two regulations are illustrated.

For example, the OSHA PSM generally qualifies as meeting the RMP prevention program element. It is important to point out that in PSM, process safety techniques employ systematic methods for evaluating a process system and identifying potential hazards. For instance, such techniques as check lists, what-if analysis, fault tree analysis, event tree analysis, and hazard and operability studies (HAZOP) can be used. These techniques, used to conduct the PHA for PSM, work well in satisfying the prevention program element requirement of RMP.

There are other complementary or similar elements shared by the two regulations. For example, both regulations share goals (1) to prevent the accidental releases of regulated substances, and (2) to minimize the consequences of releases that do occur.

At this point, the reader is probably asking the obvious question: If RMP and PSM are so similar, why do we need two different regulations? We need two different regulations because even though they are similar, they are also different in a few ways. Before getting to these differences, however, there are a few more similarities that need to be discussed.

Additional similarities (overlap) between RMP and PSM can be seen quite clearly if the facility under discussion or review is classified as a Program 3 facility. Assuming that a facility is categorized at the Program Level 3, then the following is required by the PSM:

- Process safety information
- Process hazard analysis
- Operating procedures
- Training
- Mechanical integrity
- Management of change
- Compliance audits
- Incident investigation
- Employee participation
- Hot work permit
- Contractors
- Prestartup review

It is interesting to note that facilities that are classified as Program 2 facilities do not have to include the management of change, prestartup review, employee participation, hot work permit, or contractors' elements of PSM into their RMP.

RMP/PSM: Differences

The first major difference between RMP and PSM is the source of their origination. RMP is an EPA regulation. Along with its goal to reduce the harmful effects of accidental or intentional (terrorism) spills or releases, the EPA targets protection for those entities outside the "fence line." That is, the EPA is concerned with providing protection for the public, those who do not live or work on the covered facility. PSM, an OSHA regulation, on the other hand, targets its regulatory power toward ensuring the protection to the worker, the personnel who work on the plant site. One could almost say that OSHA requires compliance inside the fence line only, as if it were a solitary, isolated entity. On the same note, one could also say that the EPA's RMP rule knocks down this fence.

This difference in philosophy of who is to be protected—the public or the worker—by a particular regulation actually works to ensure that both are protected. This is the case because facilities affected by RMP generally are also affected by the requirements of PSM. Simply stated, by complying with the requirements of each regulation, both the pubic and the worker will be protected—the environment benefits as well.

In protecting the public, the EPA requires the covered facility to conduct an OCA. In PSM, the employer is only required to investigate incidents that resulted in or could have resulted in a catastrophic release of a highly hazardous chemical in the workplace.

Other differences between RMP and PSM can be seen in the reporting requirements and in some of the different terms and definitions used by the EPA in RMP. In regards to reporting requirements, under PSM, OSHA requires the covered facility to comply with all applicable paragraphs. This compliance is expected to be completed by the covered facility, but there is no reporting requirement (i.e., submission of a formal written document showing that compliance has been effected is not required under PSM).

However, this is not the case with RMP. In addition to requiring full compliance by those facilities covered under the regulation, RMP also requires each source to submit a risk management plan.

With the exception of some key terms and phrases, the Program 3 prevention program language in RMP is identical to the OSHA PSM language. Most of the differences are in terminology based on specific legislative authorities given to the EPA or OSHA that have essentially the same meaning. Table 5.6 illustrates some of these differences.

In addition to using a few different key terms, RMP uses the following terms that are unique to the rule or are borrowed from PSM:

- Offsite—This is an area beyond the property boundary of the stationary source or an area within the property boundary to which the public has routine and unrestricted access during or outside business hours. Note: OSHA's jurisdiction includes visitors on the property who are conducting business as employees of other companies but does not necessarily extend to casual visitors or to areas within a facility boundary to which the public has routine and unrestricted access at any time.

Table 5.6. PSM and RMP Terms

PSM Term	*RMP Term*
Highly hazardous substance	Regulated substance
Employer	Owner or operator
Facility	Stationary source
Standard	Rules or part

- Process—The EPA has basically adopted OSHA's definition of *process*; the EPA decided to coordinate interpretations of the definition of process with OSHA to ensure that the rule is applied consistently.
- Significant accidental release—Any release of a regulated substance that has caused or has the potential to cause off-site consequences such as death, injury, or adverse effects to human health or the environment, or to cause the public to take shelter in place or be evacuated to avoid such consequences.
- Stationary source—The EPA defined *source* to include the entire facility. Sources are still required to submit one RMP for all processes at the source with more than a TQ of a regulated substance.

Summary of RMP Requirements

The owner or operator of a stationary source that has more than a TQ of a regulated substance in a process must complete the following:

- Prepare and submit a single RMP that covers all affected processes and chemicals.
- For Program Level 1, conduct a worst-case release scenario analysis, review accident history, and ensure emergency response procedures are in place and coordinated with community officials.
- For Program Level 2, conduct a hazard assessment, document a management system, implement a more extensive (but still streamlined) prevention program, and implement an emergency response program.
- For Program Level 3, conduct a hazard assessment, document a management system, implement a prevention program that is basically identical to the OSHA PSM Standard, and implement an emergency response program.

RMP Hazard Review Checklists, What-If Questions, and HAZOP Procedures

See figures 5.1 and 5.2 for hazards checklists.

PSM/RMP Process Hazard Analysis Methodologies

OSHA/EPA requires employers, such as the water/wastewater service sector, to perform an initial PHA on processes covered by PSM/RMP standards. The PHA must be appropriate to the complexity of the process and must identify, evaluate, and control the hazards involved in the process. Employers are required to determine and document the priority order for conducting process hazard analyses based on a rationale that takes into consideration the extent of the process hazards, number of potentially affected employees, age of the process, and operating history of the process.

HAZARD REVIEW CHECKLISTS		
General Conditions, Operations, and Maintenance	**Yes/No/NA**	**Comments**
Are work areas clean?		
Are adequate warning signs posted?		
Is ambient temperature normally comfortable?		
Is lighting sufficient for all operations?		
Are the right tools provided and used?		
Is personal protective equipment (PPE) provided and adequate?		
Are containers and tanks protected from vehicular traffic?		
Are all flammable and combustible materials kept away from containers, tanks, and feed lines?		
Are containers, tanks, and feed line areas kept free of any objects that can fall on them (e.g., ladders, shelves)? Are leak detectors with local and remote audible and visible alarms present, operable, and tested?		
Are windsocks provided in a visible location?		
Are emergency repair kits available for each type of supply present?		
Are appropriate emergency supplies and equipment present, including PPE and self-contained breathing apparatus (SCBA)?		
Are emergency numbers posted in an appropriate spot?		
Are equipment, containers, and railcars inspected daily?		
Are written operating procedures available to the operators?		
Are preventive maintenance, inspections, and testing performed as recommended by the manufacturer and industry groups and documented?		
Human Factors	**Yes/No/NA**	**Comments**
Have operators been trained on the written operating procedures and the use of PPE in normal operations (or for operators on the job		

FIGURE 5.1
Hazard review checklists for any WWTP

before June 21, 1999, have you certified that they have the required knowledge, skills, and ability to do their duties safely)?		
Do the operators follow the written operating procedures?		
Do the operators understand the applicable operating limits on temperature, pressure, flow, and level? Do the operators understand the consequences of deviations above or below applicable operating limits?		
Have operators been trained on the correct response to alarms and conditions that exceed the operating limits of the system?		
Are operators provided with enough information to diagnose alarms?		
Are controls accessible and easily understood?		
Are labels adequate on instruments and controls?		
Are all major components, valves, and piping clearly and unambiguously labeled?		
Are all components mentioned in the procedures adequately labeled?		
Are safe work practices, such as lockout/tagout, hot work, and line-opening procedures followed?		
Are personnel trained in the emergency response plan and the use of emergency kits, PPE, and cSCBAs?		
Are contractors used at the facility?		
Are contractors trained to work as safely as your own employees?		
Do you have programs to monitor that contractors are working safely?		
Chlorine and Sulfur Dioxide – Siting	**Yes/No/NA**	**Comments**
Are material safety data sheets (MSDS) readily available to those operating and maintaining the system?		
Do employees understand that there are certain materials with which chlorine/sulfur dioxide must not be mixed?		
Do employees understand the toxicity, mobility, and ability of chlorine/sulfur dioxide to sustain combustion?		

(Continues)

Do employees understand the consequences of confining liquid chlorine/sulfur dioxide without a thermal expansion device?
Do employees understand the effects of moisture on the corrosive potential of chlorine/sulfur dioxide?
Do employees understand the effects of fire and elevated temperature on the pressure of confined chlorine/sulfur dioxide and the potential for release?

Chlorine and Sulfur Dioxide – Container Shipment Unloading	**Yes/No/NA**	**Comments**
Is the truck inspected for wheel chocks, proper position, and condition of crane?		
Are adequate warning signs posted? Are there "stops"?		
Is the shipment inspected for leakage, general condition, currency of hydrostatic test, and valve protective housing before accepting?		
Are containers placed in the 6 o'clock/12 o'clock position for storage to reduce the chance of a liquid leak through the valve?		

Chlorine and Sulfur Dioxide – Bulk Shipment Unloading	**Yes/No/NA**	**Comments**
Do procedures call for hand brakes to be set and wheels chocked before unloading?		
Do procedures call for safety systems to be inspected prior to making connections for unloading or between storage tanks and transfer or distribution systems?		
For railcars, are derails to protect the open end located at least 50 feet from the car being protected?		
Are railcars staged at dead-end tacks and guarded against damage from other railcars and motor vehicles?		

FIGURE 5.1
(Continued)

Are caution signs placed at each derail and as appropriate in the vicinity of chlorine/sulfur dioxide storage, use, and transfer areas?		
Does the transfer operation incorporate an emergency shutoff system?		
Is a suitable operating platform provided at the transfer station for easy access and rapid escape?		
Is padding air for railcars from a dedicated, flow-limited, dry (to -40° F or below), and oil-free source?		
Is tank care attended as long as the car is connected, in accordance with DOT regulations?		
Building and Housing Chlorine/Sulfur Dioxide Systems	**Yes/No/NA**	**Comments**
Does the building conform with local building and fire codes and NFPA-820?		
Is the building constructed of non-combustible materials?		
Is continuous leak detection, using area chlorine/ sulfur dioxide monitors, provided in storage and process areas?		
If flammable materials are stored or used in the same building, are they separated from the chlorine/sulfur dioxide areas by a fire wall?		
Are two or more exits provided from each chlorine/sulfur dioxide storage and process area and building?		
Is the ventilation system appropriately designed for indoor operation (and scrubbing, if required) by local codes in effect at the time of construction or major modification?		
Are the exhaust ducts near floor level and the intake elevated?		
Can the exhaust fan be remotely started and stopped?		
If chlorine/sulfur dioxide are stored in the same building, are storage rooms separated as required?		
Chlorine and Sulfur Dioxide – Piping and Appurtenances	**Yes/No/NA**	**Comments**
Do piping specifications meet chlorine/sulfur dioxide requirements for the service?		

(Continues)

Do you require suppliers to provide documentation that all piping and appurtenances are certified "for chlorine service" or "for sulfur dioxide service" by the manufacturer?
Are piping systems properly supported, adequately sloped to allow drainage, and with a minimum of low spots?
Is all piping protected from all risks of excessive fire or heat?
Is an appropriate liquid expansion device or vapor pressure relief provided on every line segment or device that can be isolated?

Chlorine and Sulfur Dioxide – Design Stage of New/Modified Process	Yes/No/NA	Comments
Is the system designed to operate at lowest practical temperatures and pressures?		
If chlorine/sulfur dioxide is low enough, is the system designed to feed gaseous chlorine/sulfur dioxide from the storage container rather than liquid?		
Have the lengths of liquid chlorine/sulfur dioxide lines been minimized (reduced quantity of chlorine in lines available for release)?		
Are low-pressure alarms and automatic shutoff valves provided on chlorine/sulfur dioxide feed lines?		
Are vent-controlled spill collection sumps provided and floors sloped toward sumps for stationary tanks and railcars?		
Are vaporizers provided with automatic gas line shutoff valve, downstream pressure reducing valve, gas flow control valve, temperature control system and interlocks to shut down gas flow on low vaporizer temperature, and appropriate alarms in a continuously manned control room?		
Do vaporizers have a limited heat input capacity?		
Are curbs, sumps, and diking that minimize the surface or potential spills provided for stationary tanks and railcars?		

FIGURE 5.1
(Continued)

CHECKLIST FOR ANHYDROUS AMMONIA SYSTEMS		
Anhydrous Ammonia – Basic Rules	**Yes/No/NA**	**Comments**
Does the storage tank have a permanently attached nameplate?		
Are container(s) at least 50 feet from wells or other sources of potable water supply?		
Are container(s) painted white or other light-reflecting colors and maintained in good condition?		
Is the area free of readily ignitable materials?		
Are all main operating valves on tanks identified to show liquid or vapor service?		
Anhydrous Ammonia – Appurtenances	**Yes/No/NA**	**Comment**
Are all appurtenances designed for maximum working pressure and suitable for ammonia service?		
Do all connections to containers have shutoff valves as close to containers as practicable (except safety relief devices and gauging devices)?		
Are the excess flow and/or back pressure check valves located inside of the container or at a point outside as close as practicable to where line enters container?		
Are excess flow valves plainly and permanently marked with name of manufacturer, catalog number, and rated capacity?		
Anhydrous Ammonia – Piping	**Yes/No/NA**	**Comment**
Are piping and tubing suitable for ammonia service?		
Are provisions made for expansion, contraction, jarring, vibration, and setting?		
Is all exposed piping protected from physical damage from vehicles and other undue strain (2,000-pound pull)?		
Anhydrous Ammonia – Hoses	**Yes/No/NA**	**Comment**
Does the hose conform to TFI-RMA specifications for anhydrous ammonia?		
Is it 350 psig working, 1750 psig – burst?		
Is it marked every 5 feet with "Anhydrous Ammonia, xxx psig (maximum working pressure), manufacturer's name or trademark, year of manufacture?"		

FIGURE 5.2
Checklist for anhydrous ammonia systems

Anhydrous Ammonia – Safety Relief Devices	Yes/No/NA	Comment
Are safety relief valves installed? Are they vented upward and unobstructed to the atmosphere?		
Do they have a rain/dust cap?		
Are shutoff valves not installed between safety relief valve and container?		
Are safety relief valves marked with "NH3" or "AA," psig valve set to start-to-discharge, CFM flow at full open, manufacturer's name, and catalog number?		
Is flow capacity restricted on upstream or downstream side?		
Are hydrostatic relief valve installed between each pair of valves in liquid piping or hose?		
Anhydrous Ammonia – Safety	**Yes/No/NA**	**Comment**
Are there two suitable full face masks with ammonia canisters as approved by the Bureau of Mines? Are self-contained breathing air apparatuses required in concentrated atmospheres?		
Is an easily accessible shower or a 50 gallon drum of water available?		
Anhydrous Ammonia – Transfer or Liquid	**Yes/No/NA**	**Comment**
Are pump(s) designed for ammonia service and at least 250 psig working pressure?		
Does P.D. pump have relief valve installed ?		
Is a 0-400 psi pressure gauge installed on pump discharge?		
Are loading/unloading lines fitted with back flow check or excess flow valves?		
Are caution sign(s) posted when rail car(s) are loading/unloading?		
Are containers equipped with an approved liquid level gauging device (except those filled by weight)?		
Are containers fitted with a fixed tube liquid level gauge at 85% of water capacity?		
Anhydrous Ammonia – Stationary Tank	**Yes/No/NA**	**Comment**
Are nonrefrigerated container(s) designed for a minimum 250 (265 in CA) psig pressure?		
Are all liquid and vapor connections to container(s), except safety relief valves, liquid gauging, and pressure gauge connections, fitted with orifices not larger than No. 54 drill size equipped with excess flow valves?		

FIGURE 5.2
(Continued)

Are storage containers fitted with a 0–400 psi ammonia gauge?		
Are they equipped with vapor return valve(s)?		
Are containers marked on at least two sides with "Anhydrous Ammonia" or "Caution – Ammonia" in contrasting colors and minimum four-inch-high letters?		
Is a sign displayed stating name, address, and phone number of nearest representative, agent, or owner?		
Are containers installed on substantial concrete, masonry, or structural steel supports?		
Are ammonia systems protected from possible damage by moving vehicles?		
Anhydrous Ammonia – Basic Rules	**Yes/No/NA**	**Comments**
Are storage tank(s) painted white or other light-reflecting colors and maintained in good order?		
Is storage area free of readily ignitable materials?		
Are storage tank(s) kept away from wells or other sources of potable water supply?		
Are storage tank(s) located with ample working space all around?		
Are storage tank(s) properly vented and away from areas where operators are likely to be?		
Does receiving system include a vapor return?		
Is storage capacity adequate to receive full volume of delivery vehicle?		
Are storage tank(s) secured against overturn by wind, earthquake, and/or floatation?		
Are tank bottom(s) protected from external corrosion?		
Is aqua ammonia system protected from possible damage from moving vehicles?		
Are storage tank(s) labeled as to content?		
Are all appurtenances suitable for aqua ammonia service?		
Are all storage tank(s) fitted with liquid level gauges?		
Are liquid level gauge(s) adequately protected from physical damage?		
If tubing is used, is it fitted with a fail closed valve?		
Are all storage tank(s) fitted with overfill fittings or high level alarms?		
Are tank(s) fitted with pressure/vacuum valves?		

(Continues)

Is an ammonia gas scrubber system used?
Are piping and hose materials suitable for aqua ammonia service?
Is piping free of stain and provision made for expansion, contraction, jarring, vibration, and settling?
Is all exposed piping protected from physical damage from moving vehicles and other undue stain?
Are hoses securely clamped to hose barbs?
Are hoses inspected and renewed periodically to avoid breakage?
Are pump(s) designed for aqua ammonia service?
Are pump(s) fitted with splash guard around seals?
Are pump(s) fitted with coupling guard(s)?
Do pump(s) have local start/stop stations?
Are two (2) suitable full face masks with ammonia canisters as approved by the Bureau of Mines available?
Is a self-contained breathing air apparatus required in concentrated atmospheres?
Is an easily accessible quick acting shower with bubble fountain or 250-gallon drum of clean water available?
Is an extra pair of chemical splash proof goggles and/or full face shields available?
Is an extra set of ammonia resistant gloves, boots, coat, and apron available?
Are fire extinguishers and a first aid kit available?
Are handlers/operators wearing their goggles and gloves when working with aqua ammonia?
Is safety and first aid information posted?
Are emergency phone numbers and individuals to contact posted?

FIGURE 5.2
(Continued)

As mentioned earlier, OSHA's PSM is primarily concerned with protecting employees from the effects of accidental/intentional discharge of covered hazardous materials/chemicals—again, protecting those within the fence line. The EPA's RMP, on the other hand, is concerned with protecting those residing outside the fence line, as well as the environment. In addition to the benefits of performing a VA, a facility's overall security profile is enhanced when PSM and RMP are used in combination.

The PHA procedure can be conducted using various methodologies. For example, the checklist analysis discussed earlier is an effective technology. In addition, Pareto analysis, relative ranking, PrRA, change analysis, FMEA, fault tree analysis, event tree analysis, event and CF charting, PrHA, what-if analysis, and HAZOP can be used in conducting the PHA.

Based on experience, the what-if analysis and HAZOP seem to be the most user-friendly methodologies to use. In the following sections, the what-if analysis procedure and sample checklists used for chemicals commonly used in wastewater treatment are presented. Moreover, guide words, meanings, process parameters, and procedures for conducting HAZOP are also presented.

What-if Analysis Procedure/Sample What-if Questions

The steps in a what-if checklist analysis are as follows:

1. Select the team (personnel experienced in the process).
2. Assemble information (piping and instrumentation drawings (P&IDs), process flow diagrams (PFDs), operating procedures, equipment drawings, etc.).
3. Develop a list of what-if questions.
4. Assemble your team in a room where each team member can view the information.
5. Ask each what-if question in turn and answer the following questions: What can cause the deviation from design intent that is expressed by the question? What adverse consequences might follow? What are the existing design and procedural safeguards? Are these safeguards adequate? If these safeguards are not adequate, what additional safeguards does the team recommend? As the discussion proceeds, record the answers to these questions in tabular format.
6. Do not restrict yourself to the list of questions that you developed before the project started. The team is free to ask additional questions at any time.
7. When you have finished the what-if questions, proceed to examine the checklist. The purpose of this checklist is to ensure that the team has not forgotten anything. While you are reviewing the checklist, other what-if questions may occur to you.
8. Make sure that you follow up all recommendations and action items that arise from the hazards evaluation.

See figure 5.3 for three sample what-if checklists.

SAMPLE WHAT-IF LISTS

WHAT-IF QUESTIONS FOR CHLORINE AND SULFUR DIOXIDE SYSTEMS

1. **Movement of One-Ton Chlorine/Sulfur Dioxide Cylinders**

   What if the cylinder is dropped from the lifting apparatus?

   What if the truck rolls forward or backward?

   What if a cylinder rolls and drops from the truck?

   What if the cylinder swings while being lifted?

   What if the chlorine/sulfur dioxide container is not empty when removed from service?

   What if the automatic container switchover system fails?

   What if a chlorine/sulfur dioxide cylinder is delivered instead of sulfur dioxide/chlorine?

   What if the cylinder is not in good condition?

2. **Ton Cylinders on Trunnion, Incl. Pigtails (Subheader Lines) to Main Header Lines**

   What if pigtails rupture while connected on-line?

   What if pigtail connections open or leak when pressure is applied?

   What if something is dropped onto cylinder or connection?

   What if cracks develop in the ton cylinder flexible connection?

   What if liquid chlorine/sulfur dioxide is withdrawn through the vapor lines from the ton cylinder?

   What if the cylinder valve cannot be closed during an emergency?

   What if there are pinholes or small leaks at the fusible plugs?

   What if ton cylinder ends change shape from concave to convex?

   What if liquid is trapped between two closed valves and the temperature rises?

   What if there is a fire near the cylinders?

   What if the operator leaves the valve open and disconnects the pigtail?

   What if water enters the systems?

3. **Chlorine/Sulfur Dioxide Headers in the Chlorination (Sulfonation) Room**

   What if the pressure relief valve sticks open?

   What if a valve leaks?

   What if there is inadequate flow in the gas line (e.g., filter clogged)?

4. **Evaporators**

   What if there is overpressure in the evaporator?

   What if there is low temperature in the evaporator?

   What if rupture disks leak?

   What if the vacuum regulator valve fails?

   What if there is a gas pressure gauge leak?

   What if the vacuum regulator check unit fails?What if there is liquid chlorine/sulfur dioxide carryover to the vacuum-regulating valve downstream of the evaporator?

FIGURE 5.3
Sample what-if lists

5. **Chlorination (Sulfonation) and Pipes to Injectors**
   What if there are leaks in the chlorinator (sulfonator) unit?
   What if there is rupture of the pipe from the chlorinator to the injector?
   What if there is backflow of water into the chlorine/sulfur dioxide line?
   What if the water pump is not working?
6. **General**
   What if there is a power failure?
   What if chlorine/sulfur dioxide is released during maintenance?
   What if a chlorine/sulfur dioxide leak is not detected?
   What if there is moisture in the chlorine/sulfur dioxide system?
7. **Scrubbers**
   What if the system loses scrubber draft?
   What if the system loses scrubber solution?
   What if the manual vent to the scrubber is opened during operation?
   What if the leak tightness of the building is compromised during emergency operation of the scrubbers?
8. **Tank Trucks**
   What if the liquid hose leaks or ruptures?
   What if the vapor return hose leaks or ruptures?
   What if the truck moves?
   What if the mass of chlorine/sulfur dioxide in the truck exceeds the capacity of the tank?
   What if the chlorine tank truck is connected to a sulfur dioxide vessel (or vice versa)?
   What if there is something other than chlorine (or sulfur dioxide) in the truck?
   What if there is a fire under or near the truck?
   What if the truck collides with pipe work or building housing chlorine/sulfur dioxide storage vessels?
9. **Railcars**
   What if the liquid hose leaks or ruptures?
   What if the padding air is moist?
   What if the padding air hose ruptures?
   What if the railcar moves?
   What if the relief valve lifts below the set pressure?
   What if there is a fire under or near the truck?
   What is there is a fire on or near the railcar?

WHAT-IF QUESTIONS FOR AMMONIA SYSTEMS

1. **Storage Vessel**
   What if the vessel is overfilled?
   What if there is fire under or near the vessel?
   What if the relief valve fails to lift on demand?

(Continued)

What if the relief valve opens below its set pressure?
What if the deluge system fails to work on demand?

2. **Tank Truck Unloading**

What if the liquid unloading hose partially ruptures?
What if the liquid unloading hose completely ruptures?
What if the tank truck moves?
What if the tank truck drives away before the hose is disconnected?
What if the vapor return hose partially or completely ruptures?
What if valves are not completely closed before disconnecting the hoses?
What if the tank truck contains something other than ammonia?
What if the ammonia in the tank truck contains excess oxygen?

WHAT-IF QUESTIONS FOR DIGESTER SYSTEMS

What if something falls onto a digester cover?
What if relief valves on a digester open?
What if an intermediate digester gas storage vessel fails?
What if air is introduced into the gas collection system?
What if the gas collection header leaks or ruptures or becomes blocked?
What if a digester gas compressor fails catastrophically?
What if there is a digester gas leak into a building (digester building, compressor room, boiler room)?
What if the digester gas pressure exceeds the cover pressure rating?
What if the floating digester gas cover jams or tilts?

GENERAL WHAT-IF QUESTIONS

What if the ambient temperature is abnormally high?
What if the ambient temperature is abnormally low?
What if there is a hurricane?
What if there is a tornado?
What if there is flooding?
What if there is a heavy snowfall?
What if there is an earthquake?
What if there is a tidal wave?
What if there is a failure of electric power?

FIGURE 5.3
(Continued)

HAZOP Analysis

The HAZOP analysis technique uses a systematic process to (1) identify possible deviations from normal operations and (2) ensure that safeguards are in place to help prevent accidents. The HAZOP uses special adjectives (such as speed, flow, pressure, etc.) combined process conditions (such as *more, less, no,* etc.) to systematically consider all credible deviations from normal conditions. The adjectives, called guide words, are a unique feature of HAZOP analysis.

In this approach, each guide word is combined with relevant process parameters and applied at each point (study node, process section, or operating step) in the process that is being examined, as in table 5.7. Figure 5.4 is an example of creating deviations using guide words and process parameters.

Guide words are applied to both the more general parameters (e.g., react, mix) and the more specific parameters (e.g., pressure, temperature). With the general parameters, it is not unusual to have more than one deviation from the application of one guide word. For example, "more reaction" could mean either that a reaction takes place at a faster rate, or that a greater quantity of product results. On the other hand, some combinations of guide words and parameters will yield no sensible deviation (e.g., "as well as" with "pressure").

HAZOP Procedure

1. Select the team.
2. Assemble information (P&IDs, PFDs, operating procedures, equipment drawings, etc.).
3. Assemble your team in a room where each team member can view P&IDs.
4. Divide the system you are reviewing into nodes (you can present the nodes, or the team can choose them as you go along).

Table 5.7. HAZOP Analysis

Guide Words	*Meaning*
No	Negation of the design intent
Less	Quantitative decrease
More	Quantitative increase
Part of	Other materials present by intent
As well as	Other materials present unintentionally
Reverse	Logical opposite of the intent
Other than	Complete substitution

Common HAZOP Analysis Process Parameters

Flow	Time	Frequency	Mixing
Pressure	Composition	Viscosity	Addition
Temperature	pH	Voltage	Separation
Level	Speed	Information	Reaction

Guide Words		Parameter		Deviation
NO	+	FLOW	=	NO FLOW
MORE	+	PRESSURE	=	HIGH PRESSURE
AS WELL AS	+	ONE PHASE	=	TWO PHASE
OTHER THAN	+	OPERATION	=	MAINTENANCE
MORE	+	LEVEL	=	HIGH LEVEL

FIGURE 5.4
Example of creating deviations using guide words and process parameters

5. Apply appropriate deviations to each node. For each deviation, address the following questions: What can cause the deviation from design intent? What adverse consequences might follow? What are the existing design and procedural safeguards? Are these safeguards adequate? If these safeguards are not adequate, what does the team recommend?
6. As the discussion proceeds, record the answers to these questions in tabular format.

Final Word on PSM/RPM Compliance

Governor Tom Ridge said the following about the security role for the public professional (I interpret this to include water/wastewater professionals):

> Americans should find comfort in knowing that millions of their fellow citizens are working every day to ensure our security at every level—federal, state, county, municipal. These are dedicated professionals who are good at what they do. I've seen it up close, as Governor of Pennsylvania. . . . But there may be gaps in the system. The job of the Office of Homeland Security will be to identify those gaps and work to close them (Henry, 2002).

It is need to shore up the "gaps in the system" that has driven many water/wastewater facilities to increase security. In addition to VAs and PSM/RMP provisions, other security steps should also be taken. For example, the EPA, in its *Water Protection Task Force Alert #IV: What Wastewater Utilities Can do now to Guard against Terrorist and Security Threats* (October 24, 2001), made several recommendations to increase security and reduce threats from terrorism. It is the opinion of the author that these recommendations are the minimum steps to be taken to upgrade security. The recommendations include the following:

- Guarding against unplanned physical intrusion (water/wastewater)
  - Lock all doors and set alarms at your office, pumping stations, treatment plants, and vaults, and make it a rule that doors are locked and alarms are set.

  - Limit access to facilities and control access to pumping stations and chemical and fuel storage areas, giving close scrutiny to visitors and contractors.
  - Post guards at treatment plants, and post "Employee Only" signs in restricted areas.
  - Control access to storm sewers.
  - Secure hatches, metering vaults, manholes, and other access points to the sanitary collection system.
  - Increase lighting in parking lots, treatment bays, and other areas with limited staffing.
  - Control access to computer networks and control systems, and change the passwords frequently.
  - Do not leave keys in equipment or vehicles at any time.
- Making security a priority for employees
  - Conduct background security checks on employees at hiring and periodically thereafter.
  - Develop a security program with written plans and train employees frequently.
  - Ensure all employees are aware of communications protocols with relevant law enforcement, public health, environmental protection, and emergency response organizations.
  - Ensure that employees are fully aware of the importance of vigilance and the seriousness of breaches in security, and make note of unaccompanied strangers on the site and immediately notify designated security officers or local law enforcement agencies.
  - Consider varying the timing of operational procedures if possible in case someone is watching for patterns.
  - Upon the dismissal of an employee, change pass codes and make sure keys and access cards are returned.
  - Provide customer service staff with training and checklists of how to handle a threat if it is called in.
- Coordinating actions for effective emergency response
  - Review existing emergency response plans, and ensure they are current and relevant.
  - Make sure employees have necessary training in emergency operating procedures.
  - Develop clear protocols and chains-of-command for reporting and responding to threats; include procedures for notifying relevant emergency responders, law enforcement, environmental organizations, public health officials, consumers, and the media. Practice the emergency protocols regularly.
  - Ensure key utility personnel (both on and off duty) have access to crucial telephone numbers and contact information at all times. Keep the call list up-to-date.

- Develop close relationships with local law enforcement agencies, and make sure they know where critical assets are located. Request that they add your facilities to their routine rounds.
- Work with local industries to ensure that their pretreatment facilities are secure.
- Report to county or state health officials any illness among the employees that might be associated with wastewater contamination.
- Report criminal threats, suspicious behavior, or attacks on wastewater utilities immediately to law enforcement officials and the relevant field office of the FBI.

- Investing in security and infrastructure improvements
  - Assess the vulnerability of collection/distribution systems, water storage tanks, reservoirs, major pumping stations, water/wastewater treatment plants, chemical and fuel storage areas, outfall pipes, and other key infrastructure elements.
  - Assess the vulnerability of the stormwater collection system. Determine where large pipes run near or beneath government buildings, banks, commercial districts, and industrial facilities, or are contiguous with major communication and transportation networks.
  - Move as quickly as possible with the most obvious and cost-effective physical improvements, such as perimeter fences, security lighting, tamper-proofing manhole covers and valve boxes, and so on.
  - Improve computer systems and remote operational security.
  - Use local citizen watches.
  - Seek financing for more expensive and comprehensive system improvements.

The Bottom Line on Security

Again, when it comes to the security of water/wastewater infrastructure (and the rest of our nation), few have summed it better than Governor Ridge: "Now, obviously, the further removed we get from September 11, I think the natural tendency is to let down our guard. Unfortunately, we cannot do that. The government will continue to do everything we can to find and stop those who seek to harm us. And I believe we owe it to the American people to remind them that they must be vigilant, as well" (Henry, 2002).

REFERENCES

Clark, R. M., and Deininger, R. A. 2000. Protecting the nation's critical infrastructure: The vulnerability of U.S. water supply systems. *Journal of Contingencies and Crisis Management* 8 (2): 76–80.

Foster, S. S. D. 1987. Fundamental concepts in aquifer vulnerability, pollution risk and protection strategy. In *Vulnerability of Soil and Groundwater Pollutants*, ed. W. Van

Duijvenbooden and H. G. van Waegeningh. The Hague, The Netherlands: TNO Committee on Hydrological Research Proceedings and Information No. 38.

Henry, K. 2002. New face of security. *Gov. Security* 15 (8): 30–37.

Jeffords, J. 2005. *Wastewater Treatment Security Bill.* USC. Jim Jeffords Press Office, Vermont.

Mays, L. W. 2004. *Water Supply Systems Security.* New York: McGraw-Hill.

Minter, J. G. 1996. Prevention Chemical Accidents Still a Challenge. *Occupational Hazards* 15, no. 8, September.

Spellman, F. R. 1997. *A Guide to Compliance for PSM/RMP.* Lancaster, PA: Technomic Publishing Company.

U.S. Government Accountability Office. 2005. *Wastewater Facilities: Experts' View on How Federal Funds Should Be Spent.* GAO-05-165.

Vulnerability Assessment Fact Sheet. 2002. U.S. EPA 816-F02-025. www.epa.gov/ogwdw/security/index.html.

6

Drinking Water Contamination Threats and Incidents

Unfortunately, it appears that much of the country is of the misconception that only very large water supply systems are threatened and then really only at the downstream end, beyond the treatment system. While this situation is certainly worthy of concern, it is not a seriously realistic attack scenario.

—*U.S. Environmental Protection Agency, 2001*

Because of the seriousness of the threat of contamination to the nation's water supply, the Environmental Protection Agency (EPA) and other agencies have worked nonstop since 9/11 in gathering and providing as much advice and guidance as possible to aid water utility personnel in protecting water supplies and critical water system infrastructure. In this chapter, we provide an overview of one of the EPA's important tools in aiding water system utilities ward off and protect against the threat of water contamination. Though current material is provided, it is important to note that the toolbox is a living document—a work in progress—and is updated frequently.

CONTAMINATION THREATS AND INCIDENTS

The EPA (2003) points out that both water contamination threats and water contamination incidents could be designed to interrupt the delivery of safe water to a population, interrupt fire protection, create public panic, or cause disease or death in a population. A water contamination *threat* occurs when the introduction of a contaminant into the water system is threatened, claimed, or suggested by evidence. A water contamination *incident* occurs when a contaminant is successfully introduced into the water supply. The water contamination incident may be preceded by a threat, but not always. Both water contamination threats and incidents may be of particular concern due to the following range of potential consequences:

- Creation of an adverse impact on public health within a population
- Disruption of system operations and interruption of the supply of safe water
- Physical damage to system infrastructure
- Reduction of public confidence in the water supply
- Long-term denial of water and the cost of remediation and replacement

Keep in mind that some of these consequences would only be realized in the event of a successful contamination incident; however, the mere threat of contamination can have an adverse impact on a water system if improperly handled.

In characterizing any threat, both the possibility and probability should be considered. A general assessment of the threat of intentional contamination of drinking water indicates that it is possible to cause varying degrees of harm by contaminating a water system. Specifically, this assessment indicates the following:

- Only a few contaminants have the potential to produce widespread death or disease in a population. These contaminants include concentrated pathogens, biotoxins, and a few highly toxic chemicals that may remain stable in water long enough to adversely impact public health.
- A larger group of contaminants, including several dozen toxic chemicals, could produce localized death or disease in a segment of a population.
- Hundreds of contaminants could potentially disrupt service or undermine consumer confidence but would not result in death or disease in the population.

Note that while it is important to consider the range of possibilities associated with a particular threat, assessments are typically based on the probability of a particular occurrence. Determining probability is somewhat subjective and is often based on intelligence and previous incidents. There are historical accounts of intentional contamination of drinking water supplies with biological or chemical contaminants, but most have been associated with wartime activities. The few documented accounts of intentional contamination of a public water system in the United States have not resulted in any reported fatalities. Based on these accounts, it would appear that the probability of a successful contamination incident on a drinking water system is relatively low. However, there has been a reported increase in the interest of various terrorist groups in biological and chemical weapons. Further, some intelligence information indicates that terrorist organizations have considered water infrastructure as a possible target. Thus, the bottom line is that the potential for such an incident does exist.

While the probability of an actual contamination incident may be considered low relative to other modes of attack, the probability of the threat of contamination may be relatively high compared to other forms of attack. Many of the apparent security breaches at drinking water utilities that have occurred since 9/11 have been perceived as potential contamination incidents. Although a few threats have been verbal, most have been circumstantial, such as a low-flying airplane over a reservoir or a lock cut from the hatch of a distribution system storage tank. Given the possibility of contamination, many utilities choose to treat these security breaches as potential contamination threats. These incidents demonstrate the need for protocol to guide an appropriate response to contamination treats.

There is a common feeling among security experts that in order to prepare for contamination threats, it is necessary to generate a list of priority contaminants. However, the generation of such a list is a significant challenge due to the wide range of adverse effects that might result from intentional contamination. Further, no list of contamination threats should be considered definitive or complete. A document prepared under the auspices of the World Health Organization (WHO) succinctly sums up this dilemma in the following passage and places it in the context of planning for a response to a biological or chemical contamination incident:

> A central consideration in such preparedness planning is that it is neither possible nor necessary to specifically plan for attack by all possible biological and chemical agents. If a country is seeking to increase its preparedness to counter the effects of biological and chemical attacks, the targeting of its preparation and training on a limited but well chosen group of agents will provide the necessary capability to deal with a far wider range of possibilities. Knowledge of the general properties of this representative group of agents will enable certain measures to be taken against virtually any other agent. In addition to being impractical from a preparedness perspective, long and exhaustive lists of agents also give a misleading impression of the extent of possible threats (WHO, 2003).

Nonetheless, many federal and private organizations have generated contaminant lists that reflect the specific priorities and assumptions of that organization. While it is possible to use the experience gained from the preparation of these lists, it is very important to consider the special needs and challenges that arise from safeguarding public health through protection of the drinking water supply. For instance, there is essentially no tolerance by the public toward sudden disease and death from tainted water supplies. Another challenge is that drinking water is used not only for consumption but also for other vital uses, such as fire protection, sanitation, and industrial processes. In fact, most drinking water is used for purposes other than consumption.

CONTAMINANT CLASSES

Table 6.1 presents a number of contaminant classes that would potentially have an adverse impact if introduced into the drinking water supply. Note that this is not

Table 6.1. Contaminant Classes

Class	*Examples*	*Sources*	*Limited Access?*
	MICROBIOLOGICAL CONTAMINANTS		
Bacteria	*Bacillus anthracis* , *Brucella spp., Burkholderia spp., Campylobacter spp., Clostridium perfringens,* E. coli 0157:H7, *Francisella tularensis, Salmonella typhi, Shigella spp., Vibrio cholerae, Yersinia pestis, Yersinia enterocolitica*	Naturally occurring, microbiological labs, state-sponsored programs	Yes for select agents
Viruses	Caliciviruses, Enteroviruses, Hepatitis, A/E, Variola, VEE virus	Naturally occurring, microbiological laboratories, state-sponsored programs	Yes for select agents
Parasites	Cryptosporidium parvum, Entamoeba histolytica, Toxoplasma gondii	Naturally occurring, microbiological labs	No
	CHEMICAL CONTAMINANTS—INORGANIC		
Corrosives and caustics	Toilet bowl cleaners (hydrochloric acid), tree root dissolver (sulfuric acid), drain cleaner (sodium hydroxide)	Retail, industry	No
Cyanide salts or cyanogenics	Sodium cyanide, potassium cyanide, amygdalin, cyanogen chloride, ferricyanide salts	Supplier, industry (esp. electroplating)	Yes
Metals	Mercury, lead, osmium, their salts, organic compounds, and complexes (even those of iron, cobalt, and copper are toxic at high doses)	Industry, supplier, laboratory	Yes
Nonmetal oxyanions, organo-nonmetal	Arsenate, arsenite, selenite salts, oxyanions, organoarsenic, organoselenium compounds	Some retail, industry, supplier, laboratory	Yes
	CHEMICAL CONTAMINANTS—ORGANIC		
Fluorinated organics	Sodium trifluoroacetate (a rat poison), fluoroalcohols, fluorinated surfactants	Supplier, industry, laboratory	Yes
Hydrocarbons and their oxygenated and/or halogenated derivatives	Paint thinners, gasoline, kerosene, ketones (e.g., methyl isobutyl ketone), alcohols (e.g., methanol), ethers (e.g., methyl *vert*-butyl ether or MTBE), halohydrocarbons (e.g., dichloromethane, tetrachloroethene)	Retail, industry, laboratory, supplier	No

Insecticide	Organophosphates (e.g., Malathion), organics (e.g., DDT), carbamates (e.g., Aldicarb), some alkaloids (e.g., nicotine)	Retail, industry, supplier (varies with compound)	Yes
Malodorous, noxious, foul-tasting, and/or lachrymatory chemicals	Thiols (e.g., mercaptoacetic acid, mercapotoethanol), amines (e.g., cadavrine, putrescine), inorganic esters (e.g., trimethylphosphite, dimethyl-sulfate, acrolein)	Laboratory, supplier, police supply, military depot	Yes
Organics, water miscible	Acetone, methanol, ethylene glycol (antifreeze), phenols, detergents	Retail, industry, supplier, laboratory	No
Pesticides other than insecticides	Herbicides (e.g., chlorophenoxy or atrazine derivative), rodenticides (e.g., superwarfarins, zinc phosphide a-naphthyl thiourea)	Retail, industry, agriculture, laboratory	Yes
Pharmaceuticals	Cardiac glycosides, some alkaloids (e.g., vincristine), antineoplastic chemotherapies (e.g., aminopterin), anticoagulants, (e.g., warfarin); includes illicit drugs such as LSD, PCP, and heroin	Laboratory, supplier, pharmacy, some from a natural source	Yes
	SCHEDULE 1 CHEMICAL WARFARE AGENTS		
Schedule 1 chemical weapons	Organophosphate nerve agents (e.g., sarin, tabun, VX), vesicants, nitrogen and sulfur mustards (chlorinated alkyl amines and thioethers, respectively), Lewisite	Suppliers, military depots, some laboratories	Yes
Biologically produced toxins	Biotoxins from bacteria, plants, fungi, protists, defensive poisons in some marine or terrestrial animals (e.g., ricin, saxitoxin, botulinum toxins, T-2 mycotoxins, microcystins)	Laboratory, supplier, pharmacy, natural source, state-sponsored programs	Yes
	RADIOLOGICAL CONTAMINANTS		
Radionuclides	Does not refer to nuclear, thermonuclear, neutron bombs. Radionuclides may be used in medical devices and industrial irradiators (Cesim-137 Iridium-192, Cobalt-60, Strontium-90). Class includes both the metals and salts	Laboratory, state sources waste facilities	Yes

Source: U.S. EPA, 2003.

intended to be an exhaustive list, and there may be many others that could be used to contaminate a water supply. Note that the specific contaminants in table 6.1 do not directly correspond to the highest priority contaminants; the table is merely illustrative of the relevant contaminant classes.

In reviewing the contaminant classes listed in table 6.1, it may be apparent that many are not tightly controlled and are considered to be readily available. Most threat analysts consider availability to be the most important characteristic of a contaminant that might be used in a terrorist or criminal activity. The phrase *opportunity contaminant* has been used to describe contaminants that might be readily available even though they may be considered less than optimal from a lethality or dissemination standpoint. In many cases, specific opportunity contaminants may be more readily available on a regional or local basis. For example, a particular industrial chemical or pesticide may be produced at a facility in close proximity to the water treatment plant and its associated distribution system. Note that such site-specific considerations should be incorporated into a utility's planning and response activities, particularly with regard to threat management and analytical approaches.

To better understand the contamination threat to water, it is useful to consider other factors in addition to availability. Therefore, a broad group of potential contaminants, similar to those contained in table 6.1, were prioritized with respect to their ability to adversely impact public health. The criteria used to prioritize the contaminants are described in table 6.2. This prioritization was not intended to be comprehensive, but

Table 6.2 Criteria for Potential Water Contaminants

Criterion	*Description*
Aesthetic impacts	Changes in appearance, odor, or taste of contaminated water that might alert a consumer to the potential danger.
Availability	The ease with which the material can be obtained, synthesized, or harvested from natural sources.
Chlorine resistance	The time that a contaminant remains toxic or infectious after introduction into water containing a chlorine residual under typical distribution system conditions.
Dispersion	The ease with which a contaminant can be effectively dispersed in water.
Handling difficulty	The technical challenges associated with handling the material and introducing it into water.
Outcome of exposure	The health effects within the population resulting from exposure to the contaminant.
Potency	The amount of contaminant that would be required to contaminate a reference volume of water at a lethal or infectious dose. The smaller the amount of material, the higher the rank.
Public fear factor	Perception of the public regarding the risks associated with the contaminant.
Stability	The time that a contaminant remains toxic or infectious after introduction into an aqueous environment.
Storability	The time that a contaminant remains toxic or infectious while in storage.

Source: U.S. EPA, 2003.

rather to be inclusive of contaminant classes that warrant consideration during a contamination threat or the analysis of a water sample for an unknown contaminant.

Contamination Threat Warning Signs

A threat warning is an occurrence or discovery that indicates a potential contamination and that triggers an evaluation. It is important to note that these warnings must be evaluated in the context of typical utility activity and previous experience in order to avoid false alarms. Potential warnings include the following:

- **Security breach:** Physical security breaches, such as unsecured doors, open hatches, and unlocked/forced gates, are probably the most common threat warnings. In most cases, the security breach is likely related to lax operations or typical criminal activity such as trespassing, vandalism, and theft rather than intentional contamination of the water. However, it may be prudent to assess any security breach with respect to the possibility of contamination.
- **Witness account:** Awareness of an incident may be triggered by a witness account of suspicious activity, such as trespassing, breaking and entering, and other types of tampering. Utilities should be aware that individuals observing suspicious behavior near drinking water facilities will likely call 9-1-1 and not the water utility. In this case, the incident warning technically might come from law enforcement, as described below. Note: The witness may be a utility employee engaged in normal duties.
- **Direct notification by perpetrator:** A threat may be made directly to the water utility, either verbally or in writing. Historical incidents would indicate that verbal threats made over the phone are more likely than written threats. While the notification may be a hoax, threatening a drinking water system may be a crime under the Safety Drinking Water Act as amended by the Bioterrorism Act and should be taken seriously.
- **Notification by law enforcement:** A utility may receive notification about a contamination threat directly from law enforcement, including local, county, state, or federal agencies. As discussed previously, such a threat could be a result of suspicious activity reported to law enforcement, either by a perpetrator, a witness, or the news media. Other information, gathered through intelligence or informants, could also lead law enforcement to conclude that there may be a threat to the water supply. While law enforcement will have to lead the criminal investigation, the utility has primary responsibility for the safety of the water supply and public health. Thus, the utility's role will likely be to help law enforcement appreciate the public health implications of a particular threat as well as the technical feasibility of carrying out a particular threat.

- **Notification by news media:** A threat to contaminate the water supply might be delivered to the news media, or the media may discover a threat. A conscientious reporter would immediately report such a threat to the police, and either the reporter or the police would immediately contact the water utility. This level of professionalism would provide an opportunity for the utility to work with the media and law enforcement to assess the credibility of the threat before any broader notification is made.
- **Unusual water quality parameters:** The relationship between contamination and changes in water quality parameters is not well understood. However, it is appropriate to investigate the cause of unusual changes in water quality parameters. Changes in pH, chlorine residual, turbidity, and so on may be detected through the use of either on-line monitors or grab samples. In utility operations, this data may arise from several sources: samples collected for plan operations, routine baseline monitoring programs, and monitoring systems designed to provide early warning of changes in water quality. The results of these approaches may be used to warn of a threat. However, it is vital to consider the reliability of the results from the particular detection method or on-line monitoring system (e.g., false positives/false negatives, known interferences instrument reliability, and unusual water quality conditions associated with a known cause, such as overdosing of coagulant).
- **Consumer complaint:** An unexplained or unusually high incidence of consumer complaints about the aesthetic qualities of drinking water may indicate potential contamination. Many chemicals can impart a strong odor or taste to water, and some may discolor the water. Taste and odor complaints are quite common for water utilities, but unique taste and odor problems, particularly very unusual tastes and odor complaints clustered in a geographical area, may indicate additional problems.
- **Public health notification:** In this case, the first indication that contamination has occurred is the appearance of victims in local emergency rooms and health clinics. Utilities may therefore be notified, particularly if the cause is unknown or linked to water. An incident triggered by a public health notification is unique in that at least a segment of the population has been exposed to a harmful substance. If this agent is a chemical (including a biotoxin), then the time between exposure and onset of symptoms may be a matter of hours, and thus there is the potential that the contaminant is still present. On the other hand, the incubation period for most pathogens is several days to weeks; thus, the pathogen may have moved through the distribution system and may therefore be below detectable limits, or present only in trace quantities.

RESPONSE TO WATER CONTAMINATION THREATS

The questions addressed in this section are "Why is it necessary to respond to contamination threats at all?" and "When have I done enough?"

Why Does a Utility Need to Respond to a Contamination Threat?

As discussed, it is technically possible to introduce contaminant into a public water supply, and historical evidence suggests that the threat of contamination is indeed probable. Regardless of whether contamination is actual or threatened, both deeply impact the public health mission of water utilities. Water utilities play an essential role in providing safe and reliable drinking water supplies, preventing many problems and diseases that flourish in the absence of safe water programs. Most water utilities take their public health mission very seriously, and some are proactive in developing their plans to respond to water contamination threats. They do this often because they realize that planning for contamination events may also be beneficial in developing a more effective response to other types of emergencies.

Proper planning is a delicate process because public health measures are rarely noticed or appreciated except when they fail. Consumers are particularly upset by unsafe water because safe drinking water is often viewed as an entitlement, and indeed, it is reasonable for consumers to expect a high quality product. Public health failures during response to contamination threats often take the form of too much or too little action. The results to too little action, including no response at all, can have disastrous consequences, potentially resulting in public disease or fatalities. On the other hand, a disproportionate response to contamination threats that have not been corroborated (i.e., determined to be credible) can also have serious repercussions when otherwise safe water is unavailable. The water would not only be unavailable for human consumption, but also for sanitation, firefighting, industry, and the many other uses of public water supply. These adverse impacts must be considered when evaluating response options to a contamination threat.

Considering the potential risks of an inappropriate response to a contamination threat, it is clear that a systematic approach is needed to evaluate contamination threats. One overriding question is "When has a drinking water utility done enough?" This question may be particularly difficult to address when considering the wide range of agencies that may be involved in a threat situation. Other organizations, such as the EPA, Centers for Disease Control (CDC), law enforcement agencies, health departments, and others will each have unique obligations or interests in responding to a contamination threat.

When Is Enough, Enough?

What is a suitable and sensible response to a contamination threat? The guiding principle for responding to contamination threats is one of due diligence. As discussed above, some responses to contamination threats are warranted due to the public health implications of an actual contamination incident. However, a utility could spend a lot of time and money overresponding to every contamination threat, which would not be an effective use of resources. Furthermore, overresponse to a contamination threat carries its own adverse impacts.

Ultimately, the answer to the question of due diligence must be decided at the local level and will depend on a number of considerations. Among other factors, local authorities must decide what level of risk is reasonable in the context of a perceived threat. Careful planning is essential to developing an appropriate response to contamination threats, and in fact, one primary objective of the EPA's 2003 *Response Protocol Tool Box* (*RPTB*) is to aid users in the development of their own site-specific plans that are consistent with the needs and responsibilities of the user. Beyond planning, the *RPTB* considers a careful evaluation of any contamination threat, and an appropriate response based on the evaluation, to be the most important element of due diligence.

In the *RPTB*, the threat management process is considered in three successive stages: possible, credible, and confirmed. Thus, as the threat escalates through these three states, the actions that might be considered due diligence expand accordingly. The following paragraphs describe, in general terms, actions that might be considered due diligence at these various stages.

- **Stage 1:** Is the threat possible? If a utility is faced with a contamination threat, they should evaluate the available information to determine whether or not the threat is possible (i.e., if something could have actually happened). If the threat is possible, immediate operation response actions might be implemented, and activities such as site characterization would be initiated to collect additional information to support the next stage of the threat evaluation.
- **Stage 2:** Is the threat credible? Once a threat is considered possible, additional information will be necessary to determine if the threat is credible. The threshold at the credible stage is higher than that at the possible stage, and in general, there must be information to corroborate the threat in order for it to be considered credible. Given the higher threshold at this stage, more significant response actions might be considered, such as restrictions on public use of the water (e.g., issuance of a "do not drink" notice). Furthermore, steps should be initiated to confirm the incident and positively identify the contaminant.
- **Stage 3:** Has the incident been confirmed? Confirmation implies that definitive evidence and information have been collected to establish the presence of a harmful contaminant in the drinking water. Obviously, at this stage the concept of due diligence takes on a whole new meaning since authorities are now faced with a potential public health crisis. Response actions at this point include all steps necessary to protect public health, supply the public with an alternate source of drinking water, and begin remediation of the system.

PREPARATION

During the years that I was a consultant for various utilities on the East Coast, I performed numerous pre-OSHA audit inspections and audits of various plant Process Safety Management Standard (PSM)/Risk Management Program (RMP) compliance programs. During these site visits, one factor always seemed to be universal. While walking around the plant to gauge the plant's overall appearance and status with OSHA and EPA compliance, I almost always found that the plant managers or superintendents who accompanied me were shocked to find out what was going on in their facilities. They would scratch their heads and ask various workers, "What are you doing? Where did that new machine come from? When was it installed? Why is that door broken? Who told you to paint that door, machine, or other apparatus? When did that hole get in the fence? Who left the back gate open?" And so on.

In one inspection I performed at a plant site right after 9/11, I drove up to the gate entrance and was impressed with the height and condition of the barbed wire-topped fence and gates. I could not enter through the gates until I identified myself over a speaker system while a closed-circuit TV camera focused on me. I was let in and given instructions to sign in at the main office. I thought, not bad, just the way security should be. After walking through most of the plant site in the company of the plant manager, we approached the back fence area, which was close to the chlorine storage building. I noticed right away a large gate that was propped open with ivy growing on it—obviously, the gate had been in the wide open position for quite some time. I asked the plant manager why the gate was open. He stated that it was always open, that it led to a downhill path to a beach area below where the plant had constructed a picnic area facing the James River.

I walked the path to the bottom picnic area, looked around, and then looked back up the path toward the open gate and the prominent structure standing within, the chlorine storage building. While walking back up the path to the fence gate, I asked the plant manager if he was not concerned about the safety of the plant and especially the 50 tons of chlorine gas in the chlorine storage building.

He said, "Nah, we're safe here. I really don't see anyone swimming up river just to get into the plant site. Besides, we are surrounded by woods out here. There's nothing to attack anyway."

Once inside the plant, I asked the plant manager if he was not worried about terrorists or disgruntled former employees with a boat filled with explosives or some other weapon(s) gaining easy access to the plant and especially the chlorine building.

He said, "Nah, that will never happen. Who would be that stupid?"

Later, when I checked the Geographic Information System (GIS) system materials pertaining to the plant and surrounding area, I noted that about one mile from the plant site was a large housing area, a brewery (with 700+ employees), and a very large theme and historical park.

Know Your Water System

All utility managers and operating personnel must know their plants. There is no excuse for not knowing every square inch of the plant site. In particular, plant workers should know about any and all construction activities underway on the site, the actual construction parameters of the plant, and especially the operation of all plant unit processes. In addition, plant management must not only know its operating personnel but also its customers.

Construction and Operation

Each water system is unique with respect to age, operation, and complexity. Distribution systems are particularly unique in that many comprise a complex—and often undocumented—mix of relatively new and old components. Accordingly, understanding a distribution system as it relates to water security and response planning may be an equally complex task. Despite the challenges to understanding a water supply system, the benefits of doing so could include effectively managing threats and preventing the spread of potentially contaminated water. For instance, the water system may have structural features that enable effective isolation of a contaminated area. Also, it may become readily apparent, with full knowledge of system vulnerabilities, where and how a contaminant could be introduced.

There are ways to gain a better understanding of a particular water system, one of which is through a vulnerability assessment. Perpetrators who intentionally contaminate water may seek to produce an adverse consequence through exploitation of vulnerabilities—like open fence gates. All drinking water plants are, to some degree, vulnerable to intentional contamination incidents. The nature and extent of these vulnerabilities depends on a number of factors, such as source water type, treatment plant type, type of primary disinfectant used, residual disinfectant used in the distribution system, and security measures already in place. An assessment of the drinking water plant and system may help to identify key locations that are vulnerable to intentional contamination, or opportunity contaminants that might be more prevalent and available in the area. Better understanding of the vulnerabilities of a water system provides a basis for improving physical security against intentional contamination and preparing for the evaluation of contamination threats. Accordingly, the Bioterrorism Act established requirements that community water systems serving more than 3,300 individuals perform a system-specific vulnerability assessment for potential terrorist threats, including intentional contamination.

Another aspect of the water system that may be important, particularly in evaluating the potential spread of a suspected contaminant, is its hydraulic configuration and operation. Propagation of contaminant through a system is dependent on a number of factors, including mixing conditions at the point of contamination, hydraulic conditions within the system at the time of the contaminant introduction, and reactions between the contaminant and other materials in the system. There are several techniques for understanding the hydraulics of a water supply system. Developing this understanding may be as complex as utilizing a GIS system in conjunction with a hydraulic modeling program or as simple as manually mapping the pressure and flow zones within a system.

Information about construction materials used in the system may be contained within the utility records and can be useful in evaluating the fate and transport of a particular contaminant through a system. For example, a particular contaminant may adsorb to the pipe material used in a utility's distribution system, and this type of information could be critical in evaluating remediation options following a contamination incident.

Personnel

The employees of a water utility are generally its most valuable asset in preparing for and responding to water contamination threats and incidents. They have knowledge of the system and water quality, and may also have experience in dealing with previous contamination threats. The importance of knowledgeable and experienced personnel is highlighted by the complexity of most water treatment and distribution systems. This complexity makes the success of a contamination contingent on detailed knowledge of the system configuration, hydraulic conditions, usage patterns, and water quality. If perpetrators have somehow gained a sophisticated understanding of a water supply system, the day-to-day experience of water system personnel will prove an invaluable tool to countering any attacks. For instance, personnel may continually look for unusual aspects of daily operations that might be interpreted as a potential threat warning and may also be aware of specific characteristics of the system that make it vulnerable to contamination.

Customers

Knowledge of water system customers is an important component of preventing and managing contamination incidents. Prevention is based largely on understanding potential targets of contamination. Of special concern may be hospitals, schools, government buildings, or other institutions where large numbers of people could be directly or indirectly affected by a contamination threat or incident. Steps taken to protect the drinking water supply for these critical customers, such as enhancements to the physical security of distribution system elements at these locations, may deter the attack itself.

Water customers vary significantly with regard to their expectations of what constitutes acceptable water service, so it is necessary to consider the manner in which water is used in a particular system. For example, high water demand that is largely driven by industry has different implications compared to high usage rates in an urban center with a high population density. Some customers, such as hospitals and nursing homes, may have certain water quality requirements. Sensitive subpopulations, including children and the elderly, can exhibit adverse health effects at doses more than one order of magnitude lower than those necessary to produce disease or death in a healthy adult. That being said, for the purposes of managing water contamination threats, it is important to keep in mind that the most important goal is protecting the health of the public as a whole. Planning, preparation, and allocation of resources should be directed toward protecting the public at large, beyond specific demographic groups or individual users.

Update Emergency Response Plans for Intentional Contamination

Emergency Response Plans (ERPs) are nothing new to water utilities, since many have developed ERPs to deal with natural disasters, accidents, violence in the workplace, civil unrest, and so on. Because water utilities are a vital part of the community, it has been prudent for many utilities to develop ERPs to help ensure the continuous flow of water to the community. However many water utility ERPs developed prior to 9/11 do not explicitly deal with terrorist threats, such as intentional contamination. Recently, the U.S. Congress required community water systems serving a population greater than 3,300 to prepare or revise, as necessary, an ERP to reflect the findings of a vulnerability assessment and to address terrorist threats.

Establish Communication and Notification Strategies

Communication strategies must be planned and made available to all potential participants prior to an actual incident or threat. For the purposes of responding to a water contamination threat, the communication structure could have several management levels within the utility, as well as some external to the utility that may be involved in management of a contamination threat. Potential participants include the utility, local government, the regional government (i.e., county), state government, and federal government. Not all of these levels would necessarily be involved in every situation; however, the mechanism and process through which they interact must be decided in advance of an incident to achieve optimal public health and environmental protection. Due to the number and variety of possible participants, planning for effective communication is critical.

Perform Training and Desk/Field Exercises

In addition to a lack of planning, another reason that emergency ERPs fail is lack of training and practice. Training provides the necessary means for everyone involved to acquire the skills to fulfill their roles during an emergency. It may also provide important "buy-in" to the response process from both management and staff, which is essential to the success of any response plan. Desk exercises (also known as *tabletops* or *sand lots*), along with field exercises, allow participants to practice their skills. Also, these exercises will provide a test of the plan itself, revealing strengths and weaknesses that may be used to improve the overall plan. Improvements can include measures not only for intentional contamination of water, but also for other emergencies faced by the water utility and the community at large.

Enhance Physical Security

Denying physical access to key sites within the water system may be a deterrent to a perpetrator. Criminals often seek the easiest route of attack, just like a burglar prefers a house with an open window. Aside from deterring actual attacks, enhancing physical security has other benefits. For example, installation of fences and locks may reduce the rate of false alarms. Without surveillance equipment or locks, it may not be possible to determine whether a suspicious individual has actually entered a vulnerable area. The presence of a lock and a determination as to whether it has been cut or broken provides sound—although not definitive—evidence that an intrusion has occurred. Likewise, security cameras can be used to review security breaches and determine if the incident was simply trespassing or is a potential contamination threat. The costs of enhancing physical security may be justified by comparison to the cost of responding to just one credible contamination threat involving site characterization and lab analysis for potential contaminants. Chapter 9 provides a more in-depth discussion of physical security devices.

Establish a Baseline Monitoring Program

Background concentrations of suspected or tentatively identified contaminants may be extremely important in determining if a contamination incident has occurred. In some cases, and for some contaminants, background levels may be at detectable concentrations. Baseline occurrence information is derived from monitoring data and is used to characterize typical levels of a particular contaminant or water quality parameter. Baseline data may be used for the two following purposes in the context of emergency water sampling:

- If general water quality parameters, such as pH, chlorine residual, conductivity, or others, are used as indicators of possible contamination incidents, a baseline must be established so that significant deviations from the baseline can be observed.
- If a specific contaminant is detected in the water, knowledge of typical background levels may be necessary to properly interpret the results.

Utilize and Understand On-line Monitoring

On-line monitors are a topic of much interest, although there is a significant level of debate regarding their effectiveness as early warning systems. American Water Works Association Research Foundation (AWWARF) has published a report discussing on-line monitoring for drinking water utilities (AWWARF, 2002), which outlines the cost-benefit analysis for on-line monitoring. Many of the costs and benefits are based on issues of general water quality, plant operations, and regulatory compliance. One definite benefit is early detection of changes in water quality parameters, such as pH, chlorine residual, and turbidity. Changes in these parameters relate to treatment plant operations and may also indicate potential water contamination if properly interpreted. For instance, on-line monitoring may help establish typical background levels of the monitored parameters. These established background levels can then be compared with levels recorded during a suspected contamination incident. Another benefit of on-line monitoring for water security is that it can free operators from manual data collection and facilitate analysis and interpretation of the data for routine and security purposes. Such information should be integrated into the information management plan.

SITE CHARACTERIZATION AND SAMPLING

Site characterization is defined as the process of collecting information from an investigation site in order to support the evaluation of a drinking water contamination threat. Site characterization activities include the site evaluation, field safety screening, rapid field testing of the water, and sample collection. The investigation site is the focus of site characterization activities, and if a suspected contamination site has been identified, it will likely be designated as the primary investigation site. Additional or secondary investigation sites may be identified due to the potential spread of a suspected contaminant. The results of site characterization are of critical importance to the threat evaluation process.

There are two broad phases of site characterization: planning and implementation. The incident commander is responsible for planning, while the site characterization team is responsible for implementing the site characterization plan. This section is intended as a resource for those involved in either the planning or the implementation phases of site characterization. While the target audience is primarily drinking water utility managers and staff, other organizations (e.g., police, fire departments, FBI, and EPA criminal investigators) may be involved in site characterization activities.

SITE CHARACTERIZATION PROCESS

The EPA (2003) points out that the site characterization process is considered in the following five stages:

1. **Customizing the Site Characterization Plan:** A site characterization plan is developed for a specific threat (possibly from a generic site characterization plan) and guides the team during site characterization activities.
2. **Approaching the Site:** Before entering the site, an initial assessment of site conditions and potential hazards is conducted at the site perimeter.
3. **Characterizing the Site:** The customized site characterization plan is implemented by conducting a detailed site investigation and rapid testing of the water.
4. **Collecting Samples:** Water samples are collected in the event that lab analysis is required.
5. **Exiting the Site:** Following completion of site characterization, the site is secured and personnel exit the site and undergo any necessary decontamination.

While site characterization can be considered and implemented as a discrete process, it is important to regard it as an element of the threat evaluation process. In particular, site characterization is an activity initiated in response to a possible contamination threat in order to gather information to help determine whether or not the threat is credible. Initially, information from the threat evaluation supports the development of the customized site characterization plan. As this plan is implemented, the observations and results from site characterization feed into the threat evaluation. In turn, the revised threat evaluation may indicate that the threat is credible, not credible, or that the site characterization plan needs to be received in the field to collect more information to make this determination. Because threat evaluation and site characterization are interdependent, the incident commander must be in constant communication with the site characterization team while they are performing their tasks.

The first step is to develop a customized site characterization plan, which is based on the specific circumstances of the threat warning. This customized plan may be adapted from a generic site characterization plan, which is developed as part of a utility's preparation for responding to contamination threat. The site characterization team will use the customized plan as the basis for their activities at the investigation site. After an initial evaluation of available information, it is necessary to identify an investigation site where site characterization activities will be conducted. During the development of the customized plan, it is important to conduct an initial assessment of site hazards, which is critical to the safety of the site characterization team and may impact the makeup of the team. If there are obvious signs of hazards at the site, then teams trained in hazardous materials (hazmat) safety and handling techniques, such as a HazMat team, may need to conduct an initial hazard assessment at the site. The HazMat team will either clear the site for entry by utility personnel or decide to perform all site characterization activities themselves. Obvious signs of hazards would provide a basis for determining that a threat is credible. Furthermore, the site might be considered a crime scene if there are obvious signs of hazards, and law enforcement may take over the site investigation.

Upon arrival at the site perimeter, the team first conducts field safety screening and observes site conditions. The purpose of field safety screening activities is to identify potential environmental hazards that might pose a risk to the site characterization team. The specific field safety screening performed should be identified in the site characterization plan and might include screens for radioactivity and volatile organic chemicals. If the team detects signs of hazard, they should stop their investigation and immediately contact the incident commander to report their findings.

If no immediate hazards are identified during approach to the site, the incident commander will likely approve the team to enter the site and perform the site characterization. During this stage, the team will continue field safety screening at the site, conduct a detailed site investigation, and perform rapid field testing of the water that is suspected of being contaminated.

The team performing rapid field testing has three objectives: (1) provide additional information to support the threat evaluation process, (2) provide tentative identification of contaminants that would need to be confirmed later by lab testing, and (3) determine if hazards tentatively identified in the water require special handling precautions. The specific rapid field testing performed should be identified in the site characterization plan, and might include tests for chlorine residual and cyanide, for example. Specific field testing performed should be based on the circumstances of the specific threat and should reflect the training, experience, and resources of the site characterization team. Negative field test results are not a reason to forgo water sampling, since field testing is limited in scope and can result in false negatives.

Following rapid field testing, samples of the potential contaminated water will be collected for potential lab analysis. The decision to send samples to a lab for analysis should be based on the outcome of the threat evaluation. If the threat is determined to be credible, then samples should be delivered immediately to the lab for analysis. The analytical approach for samples collected from the site should be developed with input from the supporting lab(s), based on information from the site characterization and threat evaluation. On the other hand, if the threat is determined to be not credible, then samples should be secured and stored for a predetermined period in the event that it becomes necessary to analyze the samples at a later time.

At this point, response actions may be implemented to protect public health. However, if the threat is determined to be not credible, then samples may be collected, preserved, and stored in the event that it becomes necessary to analyze them later.

Upon completion of site characterization activities, the team should prepare to exit the site. At this stage, the team should make sure that they have documented their findings, collected all equipment and samples, and resecured the site (e.g., lock doors, hatches, and gates). If the site is considered to be a potential hazardous site or crime scene, there may be additional steps involved in exiting this site.

Roles and Responsibilities

The incident commander and the site characterization team leader are key personnel in site characterization. The incident commander has overall responsibility for managing the response to the threat, and is responsible for planning and directing site characterization activities. The incident commander may also approve the site characterization team to proceed with their activities at key decision points in the process (e.g., whether or not to enter the site following the approach).

The site characterization team leader is responsible for implementing the site characterization plan in the field and supervising site characterization personnel. The site characterization team leader must coordinate and communicate with the incident commander during site characterization.

Depending on the nature of the contamination threat, other agencies and organizations may be involved or otherwise assume some responsibility during planning and implementation of site characterization activities. Various organizations that may be involved in site characterization are described below, with their potential roles and responsibilities. The incident commander has ultimate responsibility for determining the scope of the site characterization activities and the team makeup.

PLANNING FOR SITE CHARACTERIZATION

Training staff involved in site characterization and sampling activities is critical. Responding to the site of a potential contamination incident is very different from the routine inspections and sampling activities performed by utility staff. The equipment and safety procedures used at the site of a potential contamination incident may differ significantly from those used during more typical field activities. Providing staff training in the procedures presented in this section will help to ensure that the procedures are properly and safely implemented during emergency situations.

Safety and Personnel Protection

Proper safety practices are essential for minimizing risk to the site characterization team and must be established prior to an incident to be effective. Field personnel involved in site characterization activities should have appropriate safety training to conform to appropriate regulations, such as OSHA 1910.120, which deals with hazardous substances. If planners and field personnel do not conclude that these regulations are applicable to them, they may still wish to adopt some of the safety principles in these regulations. The following guidance is provided to help users develop their own safety policies and practices. These safety policies should be consistent with the equipment and capabilities of the site characterization team and any applicable regulations.

The appropriate level of personal protection necessary to safely perform the site characterization activities will depend on the assessment of site hazards that might pose a risk to the site characterization team. An initial site hazard assessment will be performed during the development of a customized site characterization plan. The hazard assessment may be further refined during the approach to the site, based on the results of the field safety screening and initial observations of site conditions. Two general scenarios are considered, one in which there are no obvious signs of immediate hazards, and one in which there are indicators of site hazards.

Sample Collection Kits and Field Test Kits

Sample collection kits will generally contain all sample containers, materials, supplies, and forms necessary to perform sample collection activities. Field test kits contain the equipment and supplies necessary to perform field safety screening and rapid field testing of the water. Sample collection kits will generally be less expensive to construct than field test kits, and by constructing these two types of kits separately, sample collection kits can be prepositioned throughout a system while the more expensive field kits may be assigned to specific site characterization teams or personnel.

The design and construction of sample collection and field test kits is a planning activity, since these kits must be ready to go at a moment's notice in response to a possible contamination threat. In addition to improving the efficacy of the site characterization and sampling activities, advance preparation of sample collection and field test kits offers the following advantages:

- Sample collection and field test kits can be standardized throughout an area to facilitate sharing of kits in the event of an emergency that requires extensive sampling.
- Collection of a complete sample set is more likely to be achieved through the use of predesigned kits.
- Sample collection kits can be prepositioned at key locations to expedite the sampling process.
- Personnel responsible for site characterization can become familiar with the content of the kits and trained in the use of any specialized equipment.

Generic Site Characterization Plan

A site characterization plan is developed to provide direction and communication between the incident commander and the site characterization team, which will safely and efficiently implement site characterization activities. The plan should be developed expeditiously since the site characterization results are an important input to the threat evaluation process. The rapid development of a site characterization plan can be facilitated by the development of a generic site characterization plan, which is easily customized to a specific situation. While the circumstances of a particular threat warning will dictate the specifics of a customized site characterization plan, many activities and procedures will remain the same for most situations, and these common aspects can be documented in the generic site characterization plan. Potential elements of a generic plan include pre-entry criteria, communications, team organization and responsibilities, safety, field testing, sampling, and criteria for exiting the site.

Pre-entry criteria define the conditions and circumstances under which site characterization activities will be initiated and the manner in which these activities will proceed. At each stage of the process (e.g., approach to the site, onsite characterization activities, sample collection, and exiting the site), specific criteria may be defined for proceeding to the next stage. The pre-entry criteria may also specify the general makeup of the site characterization team under various circumstances. For example, under low hazard conditions, utility teams may perform site characterization, while specially trained responders might be called upon to assist in the case of potentially hazardous conditions at the site. The criteria developed for a particular utility should be consistent with the role that the utility has assumed in performing site characterization activities

The generic plan should define communication processes to ensure rapid transmittal of findings and a procedure for obtaining approval to proceed to the next stage of site characterization. It is advisable for the site characterization team to remain in constant communication with the incident commander for the entire time that they are on site. The plan should provide an approval process for the team to advance through the approach and onsite evaluation stages of the characterization to ensure that the team is not advancing into a hazardous situation. Communication devices (e.g., cell phones, two-way radios, or panic buttons) can be used to alert incident command of problems/observations encountered in the field. The communication section of the generic plan should also discuss coordination with other agencies (e.g., law enforcement, fire department) and contingencies for contacting hazmat responders.

Field testing and sampling may be handled in the generic plan by presenting a menu that covers all potential options available to the utility, based on both internal and external capabilities. In developing a customized plan, the incident commander can simply check off the field tests and sampling requirements that are appropriate for the specific situation. The site characterization plan may also need to be revised in the field based on the observations of the team.

Evaluation of Baseline Water Quality Information

Baseline water quality information is derived from routine monitoring data and used to characterize typical levels of a particular contaminant or water quality parameter. While there are no requirements to develop baseline water quality information, it can be a valuable resource when interpreting the results from site characterization and lab analysis for the following reasons:

- The results of general water quality parameters, such as pH, chlorine residual, or conductivity, among others, should be compared against a baseline to determine whether or not the results represent a significant deviation from typical levels.
- A positive result for a specific contaminant may need to be compared against typical background levels in order to properly interpret the results.

Since each of these applications of baseline data has different requirements, they are discussed separately in the following subsections.

General Water Quality Parameters

General water quality data, collected during the onsite investigation, and subsequent sample analysis may indicate water contamination if the results differ from an established baseline or typical water quality values. For such a comparison to be made, it is necessary to establish a baseline for the water quality parameter(s) of interest. Some parameters vary as a function of time and position in the system and others may experience seasonal fluctuations. These normal variations should be captured in the baseline data. The following are two approaches for establishing a general water quality baseline:

- Evaluate historical water quality monitoring data.
- During site characterization performed in response to a specific threat, baseline monitoring for target water quality parameters may be performed in an area of the distribution system that is not expected to fall within the potentially contaminated area.

Many water utilities routinely collect data that could be used to establish a baseline; however, this data would need to be analyzed and reduced to information that can be readily interpreted and used during an emergency situation. Trend charts and statistical summaries are two approaches for summarizing baseline water quality data.

In addition to using historical water quality data to establish a baseline, monitoring of unaffected sites may be used for comparison with water quality data collected from the potentially contaminated area. The unaffected site might be upstream or downstream of the potentially contaminated area, and ideally it would be hydraulically isolated from this area. However, the results of supplemental baseline monitoring must account for typical water quality variations that occur at different locations within a distribution system.

A baseline can be established for any water quality parameter that is routinely monitored. The following list of routinely monitored water quality parameters illustrates factors that may be considered when establishing a baseline:

- pH
- Conductivity
- Chlorine/chloramines residual
- Total organic carbon
- Ultraviolet absorbance

Another factor to consider when establishing a baseline for distribution system water quality is the potential for the blending of water from different treatment plants. If multiple treatment plants feed the distribution system, the water quality will be a function of the blending ratio of the water from the different plants, in addition to the other factors described above. The task of establishing a baseline for such systems is further complicated by the fact that the blending ratios will vary both spatially and temporally.

Background Levels of Specific Contaminants

The second application of summarized baseline data is supporting the interpretation of the site characterization results for a specific contaminant. If a contaminant is tentatively identified or analytically confirmed, it may be prudent to compare the results to baseline concentrations of that contaminant in the distribution system. This would be particularly important for typical water contaminants (such as cyanide, arsenic, specific disinfection byproducts, certain pesticides, *Escherichia coli*, etc.). As with general water quality parameters, there are two approaches for estimating baseline levels of a specific contaminant in a distribution system, which include the following:

- Evaluate historical monitoring results for the specific contaminant, if available.
- During site characterization performed in response to a specific threat, sampling for the specific contaminant may be performed in an area of the distribution system that is not expected to be contaminated.

In general, few contaminants of concern are monitored frequently enough to provide sufficient data to estimate a baseline. Typically, contaminants would only be monitored if required for compliance with drinking water standards or if unregulated contaminants are known to occur in the finished water and are of significant importance or interest to the utility. When such data are available, it should be compiled and summarized (e.g., using trend charts or statistical summaries) to produce information that can be used to estimated baseline occurrence in the event of an emergency. When compiling historic data, the baseline information should also identify any contaminants that are known to not occur in the finished water.

Quality Assurance for Field Testing and Sampling

Because of the diversity of potential field testing and sampling activities during the characterization, there may be no specific quality assurance (QA) activities that apply to all sampling procedures. However, the following general QA principles would apply in most cases and are consistent with the QA guidelines published by the EPA's Environmental Response Team:

- All data should be documented on field data sheets or within site logbooks.
- All instrumentation should be operated in accordance with operating instructions as supplied by the manufacturer, unless otherwise specified in the work plan. Equipment checkout and calibration activities should occur prior to site characterization and should be documented.
- Any relevant QA principles and plans specific to the particular water utility or responding organization should be observed.

Maintaining Crime Scene Integrity

The suspected contamination site that is the focus of site characterization activities could potentially become the scene of a criminal investigation. If law enforcement takes responsibility for incident command because they believe a crime has been committed, they will control the site and dictate how any additional activities, such as site characterization, are performed. In cases in which the utility is still responsible for incident command, it may still be prudent to take precautions to maintain the integrity of the potential crime scene during site characterization activities. The following are guidelines for maintaining crime scene integrity, although this should not necessarily be considered an exhaustive list:

- Substantial physical evidence of contamination might include equipment (such as pumps and hoses), or containers with residual material. Special care should be taken to avoid moving or disturbing any potential physical evidence.

- Evidence should not be handled except at the direction of the appropriate law enforcement agency. Specially trained teams from the law enforcement community are best suited (and may be jurisdictionally required) for the collection of physical evidence from a contaminated crime scene.
- The collection of physical evidence is not generally considered time sensitive; however, site characterization and sampling activities are time sensitive due to the public health implications of contaminated water. Thus, collection of water samples may precede collection of physical evidence, and care must be taken not to disturb the crime scene while performing these activities. If samples can be collected outside of the boundaries of the suspected crime scene, it may avoid concerns about the integrity of the crime scene.
- Water samples collected for the purpose of confirming/dismissing a contamination threat and identifying a contaminant could potentially be considered evidence and should be handled accordingly.
- Since the analytical results may be considered evidence as well, it is important to use a qualified lab for analytical support. If law enforcement has taken control of the situation prior to sample collection, they may require the collection of an additional sample set to be analyzed by their designated lab.
- Photographs and videos can be taken during the site characterization for use in the criminal investigation. Law enforcement should be consulted for proper handling of photographs and videos to ensure integrity of the evidence.

Maintaining crime scene integrity during site characterization is largely an awareness issue. If the site characterization team integrates the guidelines outlined above into their onsite activities, they will go a long way toward maintaining the integrity of the crime scene.

SITE CHARACTERIZATION PROTOCOL

This section lists procedures for conducting site characterization activities. A more in-depth treatment of this important subject area is provided by the EPA's *RPTB*, which is highly recommended. The site characterization protocol is divided into the following five stages:

1. **Customizing the Site Characterization Plan:** Review the initial threat evaluation, review and customize the generic site characterization plan, identify the investigation site, conduct a preliminary hazard assessment, develop a sampling approach, and form the site characterization team.
2. **Approaching the Site:** Establish the site zone, conduct field safety screening, and observe site conditions.

3. **Characterizing the Site:** Repeat field safety screening, conduct the detailed site evaluation, and perform rapid field testing of the water.
4. **Collecting Samples:** Fill sample containers, preserve samples if necessary, and initiate chain of custody.
5. **Exiting the Site:** Perform final site check, remove all equipment and samples from the site, and resecure the location.

Note that documentation of the site characterization activities and findings is an ongoing effort throughout each phase and results in a site characterization report.

Customizing the Site Characterization Plan

The first stage of the site characterization process is the customization of the generic plan developed as part of planning and preparation for responding to contamination threats. In general, the incident commander will develop the customized plan in conjunction with the site characterization team leader. The steps involved in the development of the plan include (1) performing an initial evaluation of information about the threat, (2) identifying one or more investigation sites, (3) assessing potential site hazards, (4) developing a sampling approach, and (5) assembling a site characterization team.

SITE CHARACTERIZATION REPORT

In order to provide useful information to support the threat evaluation process and the development of an analytical approach, the findings of the site characterization should be summarized in a report. This report is not intended to be a formal document but simply a concise summary of information from the site activities that can be quickly assembled within an hour or two. The recommended content of the report includes the following:

- General information about the site
- Information about potential site hazards
- Summary of observations from the site evaluation
- Field safety screening results, including any appropriate caveats on the results
- Rapid field water testing results, including any appropriate caveats on the results
- Inventory of samples collected and the sites from which they were collected
- Any other pertinent information developed during the site characterization

SAMPLE PACKAGING AND TRANSPORT

In order to perform analysis of samples beyond rapid field testing, it will be necessary to properly package the samples for transport to the appropriate labs as quickly as possible. Prompt and proper packaging and transport of samples will do the following:

- Protect the samples from increased temperatures, which can cause bacterial growth or degradation that lead to changes in composition or concentration
- Reduce the chance of sample containers leaking or breaking, which would result in loss of sample volume, loss of sample integrity, and potential exposure of personnel to hazardous substances
- Help ensure compliance with shipping regulations

ANALYTICAL GUIDE

In this section a brief of overview of the EPA's *RPTB*, Module 4: Analytical Guide is provided. The Analytical Guide is primarily intended for lab personnel and planners who would provide analytical support to a water utility in the case of a contamination threat to the water supply. The guide is intended to be a planning tool for labs that may need to provide an analytical response in the case of a contamination threat, not a "how-to" manual for use during the actual incident.

While this guide is not based expressly on regulatory requirements, it should be recognized that failure to plan for an emergency contamination incident might lead to tragic public health consequences. Accordingly, the following are this guide's:

- Describe special lab considerations for handling and processing emergency water samples suspected of contamination with a harmful substance.
- Present model approaches and procedures for analysis of water samples suspected of contamination with a known or unknown substance. These approaches and procedures are developed to take advantage of existing methodologies and infrastructure.
- Encourage planners to develop a site-specific analytical approach and lab guide that conforms to the spirit and general principles of the model approaches. Sometimes these models may represent the best and/or only way of dealing with the analytical issues involved. Frequently, they provide an example of the most comprehensive approach.

Analytical Goals for the Three Levels of the Threat Evaluation Process

The threat evaluation process is important to a lab's analytical goals, because part of providing timely, accurate results is the proper allocation of analytical resources. The allocation of lab resources to a threat will be determined by the analytical goals of the lab and the incident's credibility category. Regardless of category, it is important to remember that even if one contaminant is identified during the analysis, the presence of additional contaminants should also be investigated. The evaluation processes at each of the three stages are as follows:

- **Possible:** While speed and accuracy are necessary analytical goals for any scenario, they take on a special meaning during the evaluation of possible incidents. For the vast majority of cases, because it is unlikely that there will be an actual contaminant, it is very important to report accurate results and to not misidentify an instrumental response. These results need to be fast, but not so fast that they are inaccurate. Keep in mind that many of the decisions about water system operations will have been made before the analytical results are back from the lab.
- **Credible:** In the few credible cases, labs may receive water samples containing potentially harmful contaminants; however, the activities performed during the threat evaluation and site characterization processes should reduce the likelihood that samples containing high hazard materials reach the lab. Thus, labs should exercise due diligence to meet the goals of protecting their personnel and providing timely, accurate analytical results.
- **Confirmed:** For the rare confirmed incidents, the labs receiving the materials should be ones with specific capabilities for the contaminants (e.g., chemical, biological, or radiological), which will be suspected or known as a result of the site characterization. Environment labs will be capable of many analyses, but are prohibited from handling materials such as Schedule 1 chemical warfare agents.

Data Interpretation

Another goal of analysis is to interpret data in an appropriate manner. Often part of the interpretation involves understanding baseline concentrations. The importance of knowing baseline levels of contaminants at a location cannot be overemphasized. Establishing baseline levels not only affects the site characterization and threat management process in terms of the proper use of analytical data, but it may also serve the larger goal of creating greater public acceptance of water from a distribution system that was once contaminated. A baseline showing acceptable levels prior to any contamination incident is the benchmark to attain during the remediation process. Achieving baseline level after remediation or mitigation activities goes a long way in assuring the public that all is back to normal again. Most labs currently do not retain this information because they do not necessarily know the sampling location. However, the lab may be aware of issues regarding background data, particularly if they are asked to render an opinion on the presence of an unusual contaminant.

Role of Labs in Response to Contamination Threats

Utilities need to have confidence that labs processing emergency water samples operate according to the following guidelines:

- Apply the analytical approach presented in this module according to the circumstances of a particular incident and the needs of the client.

- Maintain facilities and implement procedures for ensuring the security and integrity of samples and analytical results that may be considered as evidence for use in prosecution.
- Receive and process emergency samples 24/7. The lab should develop an appropriate plan for staffing, sample receipt, and internal chain of custody.
- Provide results to the client in a time frame stipulated by the client. The lab should be prepared to provide the client with an estimate of the time frame in which results may be available. The utility may need to take certain response measures before analytical results are available. Accordingly, the time frame may be dictated by site-specific factors, such as the hydraulic residency time within a segment of a distribution system.
- Implement appropriate QA and QC (quality control) procedures, and report the QC data along with the analytical results.
- Use proper channels for reporting results.
- Provide support in the analysis and interpretation of analytical results.
- Have a back-up plan for processing samples should the lab's facility become unusable or unavailable.

Labs may wish to develop their lab guide in accordance with the EPA guidelines, and to share these plans with their clients. Timely and accurate results from the lab may provide valuable input for making decisions about how to proceed with a response to a contamination threat. Identification of a harmful contaminant in a water sample would likely trigger additional public health measures, including additional sampling to characterize the spread of the contaminant, and possibly some initial remediation efforts. Likewise, if lab results reveal nothing out of the ordinary, the response would likely be terminated, and any precautionary public health measures could be cancelled or scaled down.

LABORATORY INFRASTRUCTURE IN THE UNITED STATES

The following two sections provide a general description of the lab infrastructure for the analysis of chemical and microbiological contaminants in a water matrix. The Laboratory Compendium is a comprehensive, web-based, searchable database of lab capability for environmental analysis in water, air, soil. Sediments and other media will be provided in the Laboratory Compendium when available. It is not a listing of labs approved, certified, or recommended to analyze samples from intentional contamination incidents. Rather, the compendium is designed to be a tool for searching for labs and determining their ability to perform various analytical techniques.

Chemical Laboratories

In addition to labs within water utilities, standard and specialized chemistry labs within federal, state, local, city, and municipal government agencies, as well as commercial labs, may support analysis of chemicals in water samples. Some lab resources may also be available from the academic and industrial sectors. For example, a major chemical manufacturer might want to bring their lab operations to bear during the evaluation of an incident in which their products are suspected in a water contamination threat. Many academic and industrial labs, however, may not necessarily be set up to rapidly respond to water contamination threats without extensive planning. Regardless of the lab's origin, it is anticipated that four different types of analytical chemistry labs would play a role in implementing the chemistry procedures in the analytical approach: environmental chemistry, radiochemistry, biotoxins, and chemical weapons.

Environmental Chemistry Labs

This group forms the largest sector of the laboratory infrastructure for analysis of chemicals in water and includes many USEPA, state, and commercial water analysis labs. Environmental chemistry labs are typically set up to perform analysis of water samples for compliance with the Safe Drinking Water or Clean Water Acts, as well as other chemical parameters that are important to system operations and overall water quality. It is important to realize that these labs are typically involved in the analysis of contaminants at concentrations associated with chronic (long-term) toxicity, not the acutely (short-term) toxic levels potentially associated with an intentional contamination incident. While it may seem intuitive that labs capable of determining contaminants at low concentrations should not experience difficulties at high concentrations, this is not necessarily the case for a number of technical and practical reasons.

These labs typically have the instrumentation necessary to implement standard methods for chemical analysis in a water matrix. Because many of these labs are involved in regulatory compliance, the lab and staff may already be accredited and certified to implement these methods. However, unless the lab tests for the particular chemical analyte on a routine basis, the lab will not necessarily be able to run the associated method without advance notice. This includes maintaining an inventory of standards and reagents, setting up the instrument for a particular method, and having staff trained to run the method. This may be particularly relevant in the context of chemical analysis in the case of a suspected intentional contamination incident since many of the chemicals of greatest concern are not routinely tested for in water, even though standardized methods are available.

Some environmental chemistry labs may have unique capabilities for analysis of select radionuclides or biotoxins, but this is not the expected norm. Analysis for these chemicals may need to be performed by a specialty lab. Further, many research labs are typically involved in method development, and thus are equipped with advanced equipment and highly trained analysts. These assets provide capability for exploratory techniques that are currently beyond the means of other environmental chemistry labs.

Radiochemistry Labs

If a radioactive contaminant is suspected, analyses should be performed by a lab specifically equipped to handle such material and analyze for a range of radionuclides. The EPA, Department of Energy, states, and commercial firms have labs dedicated to the analysis of radioactive and/or nuclear material. For further information about the EPA's lab service and radiological emergency response programs, contact EPA's regional office.

Biotoxin Labs

Currently, few labs are set up specifically for the analysis of biotoxins. Those in existence primarily focus on the analysis of marine biotoxins in coastal waters and seafood products. Some Laboratory Response Network (LRN) laboratories may have the capability to analyze for select biotoxins in water samples, assuming proper sample preparation. In addition, there are a number of labs in government and academia that perform biotoxin analysis, usually for other matrices than water (e.g., seafood or agricultural products). It is possible that some biotoxin analyses could be performed in qualified environmental chemistry labs using techniques such as gas chromatography/mass spectrometry, high performance liquid chromatography, and immunoassay; however, such capabilities may not currently be widespread.

Chemical Weapon Labs

For the purposes of the EPA's RPTB, a *chemical weapon* refers to those chemicals that the Chemical Weapons Convention (CWC) has placed on its Schedule 1. This list includes toxic chemicals with few or no legitimate purposes and that were developed or used primarily for military purposes. CWC monitors chemicals on two other schedules, as well as certain *unscheduled discrete organic chemicals.*

Microbiological Laboratories

The analysis of waterborne pathogens will likely be performed by either an environmental microbiology lab or a lab that is part of the LRN. This may include hospital labs, medical labs, public health labs, and environmental microbiology labs. However, the missions and capabilities of these two distinct sets of labs are significantly different, and neither may be particularly well prepared for the analysis of all biological terrorism (BT) contaminants of concern in a water matrix. The potential role of each of these types of labs in responding to a water contamination threat involving pathogens is discussed in the following two sections.

Laboratory Response Network

The LRN was developed by the Centers for Disease Control (CDC), the Association of Public Health Laboratories (APHL), and the FBI for the express purpose of dealing with BT threats, including pathogens and some biotoxins. Various labs within each state participate in the LRN. Labs that are part of the LRN can analyze the select agents subject to legislative requirements set forth in the Select Agent Regulation (42 CFR 72). The legislation requires that, subject to certain exemptions, entities possessing biological agents that are listed as select agents must register with CDC and/or the U.S. Department of Agriculture's Animal and Plant Inspection Service and demonstrate compliance with specific safety and security standards for handling these agents.

The LRN membership is organized into sentinel labs, reference labs, and national labs. Sentinel labs recognize an agent, rule it out, and/or refer the sample to reference labs for the next level for confirmatory testing. At the top of the pyramid are national labs (namely, CDC and U.S. Army Medical Research Institute of Infectious Disease), which are capable of definitive characterization of even the most hazardous biological agents.

Environmental Microbiological Labs

Environmental microbiological labs, including those of the EPA, state governmental agencies, and the commercial sector, typically perform analyses for waterborne pathogens. Most of these labs have the equipment and staff necessary to perform classical microbiological methods, and routinely analyze for indicators of fecal contamination such as fecal and total coliforms and *E. coli*. Culture techniques are available for many of the more common waterborne pathogens such as *Vibrio cholerae*, *Salmonella enteriditis*, *Typhi*, and *Shigella spp.*; however, analyses for these pathogens are not routinely performed in most environmental microbiological labs. While some environmental microbiological labs have expanded capabilities to analyze for parasites such as *Cryptosporidium* and *Giardia* or to perform molecular assays for some organisms, these capabilities are not widespread.

While many environmental microbiological labs are well equipped to analyze for microbiological contaminants in a water matrix, they generally lack the infrastructure, training, and methods to analyze for many pathogens of concern. Furthermore, only the most currently registered labs reside in the LRN. Thus, even if environmental microbiological labs develop additional capabilities for pathogen analysis, they could not perform such analyses without registering for select agents.

Integration of Lab Resources

The above sections presented a brief overview of the laboratory infrastructure that will likely be called on to implement the procedures presented in this module. While the core infrastructure may exist for both chemical and microbiological analysis, no mechanism currently exists to provide coordination in a manner conducive to optimal analytical response. At a minimum, this will create a greater logistical burden on the organization coordinating sampling and shipment to qualified labs. In the worst case, these inefficiencies may result in an incomplete analysis of an unknown substance, shipment to the wrong lab, or delays in receiving time-sensitive information.

The formation of environmental lab response networks would help to address these coordination issues. Such networks are in existence for the analysis of clinical samples (CDC's LRN) and food samples (the Food and Drug Administration's [FDA] Food Emergency Response Network [FERN]). The FERN was developed through integration with the existing LRN for pathogen analysis and establishment of regional forensic chemistry labs that serve as reference labs for other FDA labs. In the absence of a formal network, the analytical response to water contamination threats may be supported by the lab infrastructure as it currently exists.

Accordingly, the approaches described above were developed for implementation by the existing laboratory infrastructure. Some states have established or will establish network-like entities to coordinate lab efforts. The following steps may help states better integrate lab resources and provide a more coordinated response to water contamination threats:

- Establish environmental labs that are capable of implementing both basic and expanded screens for unknown chemicals in water samples.
- Establish environmental microbiology labs within the LRN that are capable of performing sentinel testing for pathogens of concern in a water matrix.
- Determine those biotoxins that will likely be analyzed for in environmental chemistry labs and those that will be analyzed for in LRN labs. Considering the range of techniques used to measure biotoxins, there may be some overlap in biotoxin capability.
- Establish a clear sample referral system for the analytical confirmation of tentatively identified contaminants in cases where the environmental chemistry lab cannot perform it. This concept is integrated into the LRN for microbiological analysis but is not formally defined for chemical analyses.

REFERENCES

American Water Works Association Research Foundation. 2002. *Online Monitoring for Drinking Water Utilities*, ed. Erika Hargesheimer. Denver, CO: American Water Works Association.

U.S. Environmental Protection Agency. 2001. *Protecting the Nation's Water Supplies from Terrorist Attack: Frequently Asked Questions.* Washington, DC: U.S. Environmental Protection Agency.

U.S. Environmental Protection Agency. 2003. *Response Protocol Toolbox: Planning for and Responding to Drinking Water Contamination Threats and Incidents.* Washington, DC: U.S. Environmental Protection Agency.

WHO. 2003. *Public Health Response to Biological and Chemical Weapons: WHO Guidance.* 2nd ed. Geneva: World Health Organization.

7

Cyber Security: SCADA

On April 23, 2000, police in Queensland, Australia stopped a car on the road and found a stolen computer and radio inside. Using commercially available technology, a disgruntled former employee had turned his vehicle into a pirate command center of sewage treatment along Australia's Sunshine Coast. The former employee's arrest solved a mystery that had troubled the Maroochy Shire wastewater system for two months. Somehow the system was leaking hundreds of thousands of gallons of putrid sewage into parks, rivers and the manicured grounds of a Hyatt Regency hotel—marine life died, the creek water turned black and the stench was unbearable for residents. Until the former employee's capture—during his 46th successful intrusion—the utility's managers did not know why.

Specialists study this case of cyber-terrorism because it is the only one known in which someone used a digital control system deliberately to cause harm. The former employee's intrusion shows how easy it is to break in—and how restrained he was with his power.

To sabotage the system, the former employee set the software on his laptop to identify itself as a pumping station, and then suppressed all alarms. The former employee was the "central control station" during his intrusions, with unlimited command of 300 SCADA nodes governing sewage and drinking water alike.

The bottom line: as serious as the former employee's intrusions were they pale in comparison with what he could have done to the fresh water system—he could have done anything he liked.

—Barton Gellman, 2002

In 2000, the FBI identified and listed threats to critical infrastructure. These threats are listed and described in table 7.1.

WASTE/WASTEWATER AND CYBERSPACE

In the past few years, especially since 9/11, it has become somewhat routine for us to pick up a newspaper or magazine or view a television news program where a major

Table 7.1. Threats to Critical Infrastructure Observed by the FBI

Threat	*Description*
Criminal groups	There is an increased use of cyber intrusions by criminal group who attack systems for purposes of monetary gain.
Foreign intelligence services	Foreign intelligence services use cyber tools as part of their information gathering and espionage activities.
Hackers	Hackers sometimes crack into networks for the thrill of the challenge or for bragging rights in the hacker community. While remote cracking once required a fair amount of skill or computer knowledge, hackers can now download attack scripts and protocols from the Internet and launch them against victim sites. Thus, while attack tools have become more sophisticated, they have also become easier to use.
Hacktivists	Hacktivism refers to politically motivated attacks on publicly accessible web pages or email servers. These groups and individuals overload email servers and hack into websites to send a political message.
Information warfare	Several nations are aggressively working to develop information warfare doctrine, programs, and capabilities. Such capabilities enable a single entity to have a significant and serious impact by disrupting the supply, communications, and economic infrastructures that support military power—impacts that, according to the director of Central Intelligence, can affect the daily lives of Americans across the country.
Inside threat	The disgruntled organization insider is a principal source of computer crimes. Insiders may not need a great deal of knowledge about computer intrusions because their knowledge of a victim system often allows them to gain unrestricted access to cause damage to the system or to steal system data. The insider threat also includes outsourcing vendors.
Virus writers	Virus writers are posing an increasingly serious threat. Several destructive computer viruses and worms have harmed files and hard drives, including the Melissa Macro Virus, the Explore.Zip worm, the CIH (Chernobyl) Virus, Nimda, and Code Red.

Source: FBI, 2000.

topic of discussion is cyber security or the lack thereof. Many of the cyber intrusion incidents we read or hear about have added new terms or new uses for old terms to our vocabulary. For example, old terms such as Trojan horse, worms, and viruses have taken on new connotations in regards to cyber security issues. Relatively new terms such as scanners, Windows NT hacking tools, ICQ hacking tools, mail bombs, sniffers, logic bomb, nukers, dots, backdoor Trojan, key loggers, hackers' Swiss knife, password crackers, and BIOS crackers are now commonly read or heard.

Not all relatively new and universally recognizable cyber terms have sinister connotation or meaning, of course. Consider, for example, the following digital terms: backup, binary, bit byte, CD-ROM, CPU, database, email, HTML, icon, memory, cyberspace, modem, monitor, network, RAM, Wi-Fi (wireless fidelity), record, software,

World Wide Web—none of these terms normally generate thoughts of terrorism in most of us.

There is, however, one digital term, SCADA, that most people have not heard of. This is not the case, however, with those who work with the nation's critical infrastructure, including water/wastewater. SCADA, or Supervisory Control and Data Acquisition System (also sometimes referred to as digital control systems or process control systems), plays an important role in computer-based control systems. Many water/wastewater systems use computer-based systems to remotely control sensitive processes and physical processes previously controlled manually. These systems (commonly known as SCADA) allow a water/wastewater utility to collect data from sensors and control equipment located at remote sites. Common water/wastewater system sensors measure elements such as fluid level, temperature, pressure, water purity, water clarity, and pipeline flow rates. Common water/wastewater system equipment includes valves, pumps, and mixers for mixing chemicals in the water supply.

WHAT IS SCADA?

Simply, SCADA is a computer-based system that remotely controls processes previously controlled manually. SCADA allows an operator using a central computer to supervise (control and monitor) multiple networked computers at remote locations. Each remote computer can control mechanical processes (pumps, valves, etc.) and collect data from sensors at its remote location.The central computer is called the master terminal unit, or MTU. The operator interfaces with the MTU using software called Human Machine Interface, or HMI. The remote computer is called a program logic controller (PLC) or remote terminal unit (RTU). The RTU activates a relay (or switch) that turns mechanical equipment on and off. The RTU also collects data from sensors.

In the initial stages utilities ran wires, also know as hardwire or land lines, from the central computer (MTU) to the remote computers (RTUs). Since remote locations can be hundreds of miles from the central location, utilities begun to use public phone lines and modems, telephone company lines, and radio and microwave communication. More recently, they have also begun to use satellite links, Internet, and newly developed wireless technologies.

Since the SCADA systems' sensors provided valuable information, many utilities established connections between their SCADA systems and their business system. This allowed utility management and other staff access to valuable statistics, such as water usage. When utilities later connected their systems to the Internet, they were able to provide stakeholders with water/wastewater statistics on the utility web pages. Figure 7.1 provides a basic illustration of a representative SCADA network. Note that firewall protection would normally be placed between Internet and business system and between business system and the MTU.

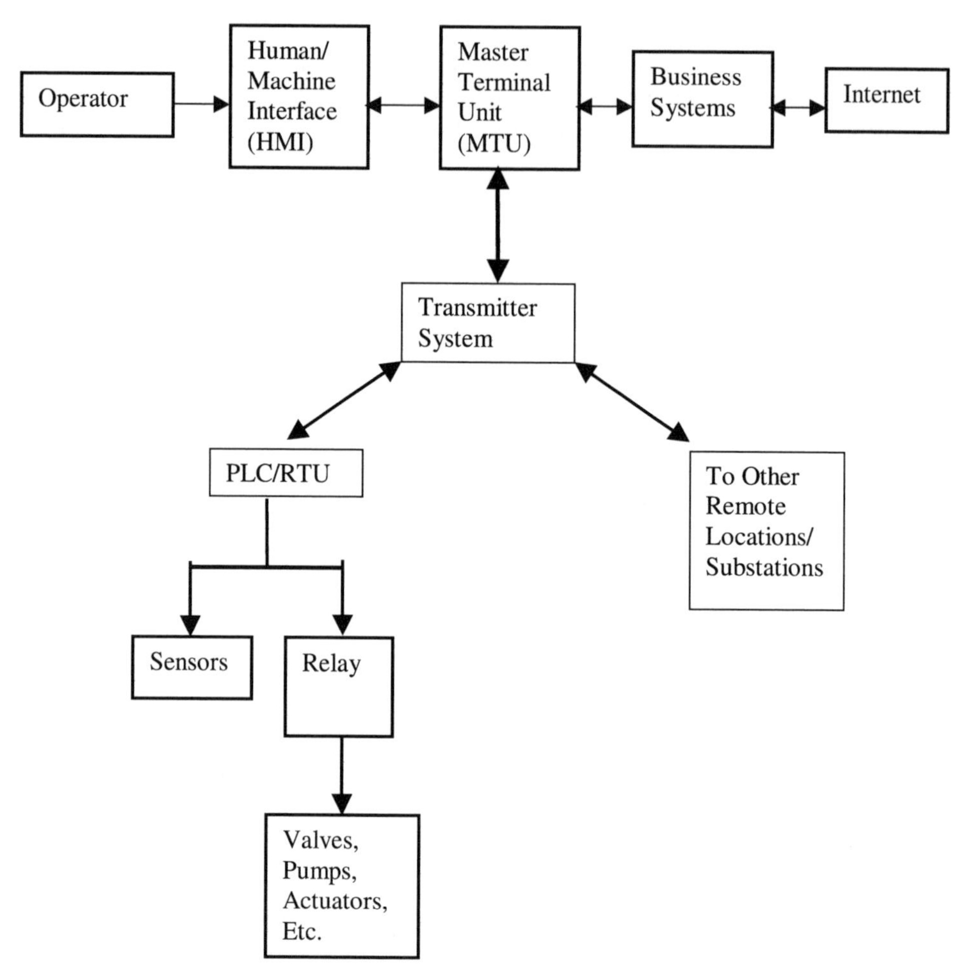

FIGURE 7.1
Representative SCADA network

SCADA Applications in Water/Wastewater Systems

As stated above, SCADA systems can be designed to measure a variety of equipment operating conditions and parameters, volumes and flow rates, or water quality parameters; they can also respond to changes in those parameters, either by alerting operators or by modifying system operations through a feedback loop system. This saves personnel from having to physically check each process or piece of equipment daily. SCADA systems can also be used to automate certain functions so that they can be performed without initiation by an operator (e.g., injecting chlorine in response to periodic low chlorine levels in a distribution system, or turning on a pump in response to low water levels in a storage tank). As described above, in addition to process equipment, SCADA systems can also integrate specific security equipment, such as cameras, motion sensors, lights, data from card-reading systems, and more, thereby providing a clear picture of what is happening throughout a facility. Finally, SCADA systems also provide constant, real-time data on processes, equipment, location access, and so on, and quickly responds to the data when necessary. This can be extremely useful during emergency conditions, such as when distribution mains break or when potentially disruptive biochemical oxygen demand spikes appear in wastewater influent.

Because these systems can monitor multiple processes, equipment, and infrastructure and then provide quick notification of, or response to, problems or upsets, SCADA systems typically provide the first line of detection for atypical or abnormal conditions. For example, a SCADA system might be connected to sensors that measure certain water quality parameters and react when those parameters are measured outside of a specific range. Or a real-time customized operator interface screen could display and control critical systems monitoring parameters.

The system could transmit warning signals back to the operators, such as by initiating a call to a personal pager. This might allow the operators to initiate actions to prevent contamination and disruption of the water supply. Further automation of the system could ensure that the system initiated measures to rectify the problem. Preprogrammed control functions (e.g., shutting a valve, controlling flow, increasing chlorination, or adding other chemicals) can be triggered and operated based on SCADA utility.

SCADA VULNERABILITIES

According to the Environmental Protection Agency (EPA, 2005), SCADA networks developed with little attention paid to security, often making the security of these systems weak. Studies have found that, while technological advancements introduced vulnerabilities, many water/wastewater utilities have spent little time securing their SCADA networks. As a result, many SCADA networks may be susceptible to attacks and misuse.

Remote monitoring and supervisory control of processes started to develop in the early 1960s. The advent of minicomputers made it possible to automate a vast number of once manually operated switches. Advancements in radio technology reduced the communication costs associated with installing and maintaining buried cable in remote areas. SCADA systems continued to adopt new communication methods, including satellite and cellular. As the price of computers and communications dropped, it became economically feasible to distribute operations and expand SCADA networks to include even smaller facilities.

Advances in information technology and the necessity of improved efficiency have resulted in increasingly automated and interlinked infrastructures. These advances also created new vulnerabilities due to equipment failure, human error, weather and other natural causes, and physical and cyber attacks. Some areas and examples of possible SCADA vulnerabilities include the following:

- **Human:** People can be tricked or corrupted, and may commit errors.
- **Communications:** Message can be fabricated, intercepted, changed, deleted, or blocked.
- **Hardware:** Security features are not easily adapted to small self-contained units with limited power supplies.
- **Physical:** Intruders can break into a facility to steal or damage SCADA equipment.
- **Natural:** Tornadoes, floods, earthquakes, and other natural disasters can damage equipment and connections.
- **Software:** Programs can be poorly written.

A study included a survey that found many water utilities doing little to secure their SCADA network vulnerabilities (Ezell, 1998). For example, many respondents reported that they had remote access, which can allow an unauthorized person to access the system without being physically present. More than 60 percent of the respondents believed that their systems were not safe from unauthorized access and use. Twenty percent of the respondents even reported known attempts, successful unauthorized access, or use of their system. Yet 22 of 43 respondents reported that they do not spend any time ensuring their network is safe, and 18 of 43 respondents reported that they spend less than ten percent of their time ensuring network safety.

The Computer Security Institute and FBI conduct an annual Computer Crime and Security Survey (FBI, 2004). The survey reported on ten types of attacks or misuse and reported that viruses and denial of service had the greatest negative economic impact. The same study also found that 15 percent of the respondents reported abuse of wireless networks, which can be a SCADA component. On average, respondents from all sectors did not believe that their organization invested enough in security awareness. Utilities as a group reported a lower average computer security expenditure/investment per employee than many other sectors, such as transportation, telecommunications, and finance.

Sandia National Laboratories' *Common Vulnerabilities in Critical Infrastructure Control Systems* describes some of the common problems it has identified in the following five categories (Stamp et al., 2003):

1. **System Data:** Important data attributes for security include availability, authenticity, integrity, and confidentiality. Data should be categorized according to its sensitivity, and ownership and responsibility must be assigned. However, SCADA data is

often not classified at all, making it difficult to identify where security precautions are appropriate.

2. **Security Administration:** Vulnerabilities emerge because many systems lack a properly structured security policy, equipment and system implementation guides, configuration management, training, and enforcement and compliance auditing.
3. **Architecture:** Many common practices negatively affect SCADA security. For example, while it is convenient to use SCADA capabilities for other purposes, such as fire and security systems, these practices create single points of failure. Also, the connection of SCADA networks to other automation systems and business networks introduces multiple entry points for potential adversaries.
4. **Network (including communication links):** Legacy systems' hardware and software have very limited security capabilities, and the vulnerabilities of contemporary systems (based on modern information technology) are publicized. Wireless and shared links are susceptible to eavesdropping and data manipulation.
5. **Platforms:** Many platform vulnerabilities exist, including default configurations retained, poor password practices, shared accounts, inadequate protection for hardware, and nonexistent security monitoring controls. In most cases, important security patches are not installed, often due to concern about negatively impacting system operation; in some cases technicians are contractually forbidden from updating systems by their vendor agreements.

The following incident helps to illustrate some of the risks associated with SCADA vulnerabilities.

While conducting a vulnerability assessment, a contractor stated that personnel from his company penetrated the information system of a utility within minutes. Contractor personnel drove to a remote substation and noticed a wireless network antenna. Without leaving their vehicle, they plugged in their wireless radios and connected to the network within five minutes. Within 20 minutes they had mapped the network, including SCADA equipment, and accessed the business network and data. This illustrates what a cyber security advisor from Sandia National Laboratories specializing in SCADA stated: Utilities are moving to wireless communication without understanding the added risks.

The Increasing Risk

According to the U.S. General Accounting Office (2003), historically, security concerns about control systems (SCADA included) were related primarily to protecting against physical attack and misuse of refining and processing sites or distribution and holding facilities. However, more recently there has been a growing recognition that control systems are now vulnerable to cyber attacks from numerous sources, including hotel governments, terrorist groups, disgruntled employees, and other malicious intruders.

In addition to control system vulnerabilities mentioned earlier, several factors have contributed to the escalation of risk to control systems, including (1) the adoption of standardized technologies with known vulnerabilities, (2) the connectivity of control systems to other networks, (3) constraints on the implementation of existing security technologies and practices, (4) insecure remote connections, and (5) the widespread availability of technical information about control systems.

Adoption of Technologies with Known Vulnerabilities

When a technology is not well known, widely used, understood, or publicized, it is difficult to penetrate it and thus disable it. Historically, proprietary hardware, software, and network protocols made it difficult to understand how control systems operated—and therefore how to hack into them. Today, however, to reduce costs and improve performance, organizations have been transitioning from proprietary systems to less expensive, standardized technologies such as Microsoft's Windows and Unix-like operating systems and the common networking protocols used by the Internet. These widely used standardized technologies have commonly known vulnerabilities, and sophisticated and effective exploitation tools are widely available and relatively easy to use. As a consequence, both the number of people with the knowledge to wage attacks and the number of systems subject to attack have increased. Also, common communication protocols and the emerging use of extensible markup language (commonly referred to as XML) can make it easier for a hacker to interpret the content of communications among the components of a control system.

Control systems are often connected to other networks—enterprises often integrate their control system with their enterprise networks. This increased connectivity has significant advantages, including providing decision makers with access to real-time information and allowing engineers to monitor and control the process control system from different points on the enterprise network. In addition, the enterprise networks are often connected to the networks of strategic partners and to the Internet. Further, control systems are increasingly using wide-area networks and the Internet to transmit data to their remote or local stations and individual devices. This convergence of control networks with public and enterprise networks potentially exposes the control systems to additional security vulnerabilities. Unless appropriate security controls are deployed in the enterprise network and the control system network, breaches in enterprise security can affect the operation of control systems.

According to industry experts, the use of existing security technologies, as well as strong user authentication and patch management practices, are generally not implemented in control systems because control systems operate in real time, typically are not designed with cybersecurity in mind, and usually have limited processing capabilities.

Existing security technologies such as authorization, authentication, encryption, intrusion detection, and filtering of network traffic and communications require more bandwidth, processing power, and memory than control system components typically have. Because controller stations are generally designed to do specific tasks, they use low-cost, resource-constrained microprocessors. In fact, some devices in the electrical industry still use the Intel 8088 processor, introduced in 1978. Consequently, it is difficult to install existing security technologies without seriously degrading the performance of the control system.

Further, complex passwords and other strong password practices are not always used to prevent unauthorized access to control systems, in part because this could hinder a rapid response to safety procedures during an emergency. As a result, according to experts, weak passwords that are easy to guess, shared, and changed infrequently are reportedly common in control systems, as is the use of default passwords or even no password at all.

In addition, although modern control systems are based on standard operating systems, they are typically customized to support control system applications. Consequently, vendor-provided software patches are generally either incompatible or cannot be implemented without service shutting down "always-on" systems or affecting interdependent operations.

Potential vulnerabilities in control systems are exacerbated by insecure connections. Organizations often leave access links—such as dial-up modems to equipment and control information—open for remote diagnostics, maintenance, and examination of system status. Such links may not be protected with authentication of encryption, which increases the risk that hackers could use these insecure connections to break into remotely controlled systems. Also, control systems often use wireless communications systems, which are especially vulnerable to attack, or leased lines that pass through commercial telecommunications facilities. Without encrypting protected data as it flows through these insecure connections or authenticating mechanisms to limit access, there is limited protection for the integrity of the information being transmitted.

Public information about infrastructures and control systems is available to potential hackers and intruders. The availability of this infrastructure and vulnerability data was demonstrated earlier this year by a university graduate student, whose dissertation reportedly mapped every business and industrial sector in the American economy to the fiber-optic network that connects them—using material that was available publicly on the Internet, none of which was classified. Many of the electric utility officials who were interviewed for the National Security Telecommunications Advisory Committee's Information Assurance Task Force's Electric Power Risk Assessment expressed concern over the amount of information about their infrastructure that is readily available to the public.

In the electric power industry, open sources of information—such as product data and educational videotapes from engineering associations—can be used to understand the basics of the electrical grid. Other publicly available information—including filings of the Federal Energy Regulatory Commission, industry publications, maps, and material available on the Internet—is sufficient to allow someone to identify the most heavily loaded transmission lines and the most critical substations in the power grid.

In addition, significant information on control systems is publicly available—including design and maintenance documents, technical standards for the interconnection of control systems and RTUs, and standards for communication among control devices—all of which could assist hackers in understanding the systems and how to attack them. Moreover, there are numerous former employees, vendors, support contractors, and other end users of the same equipment worldwide with inside knowledge of the operation of control systems.

Cyber Threats to Control Systems

There is a general consensus—and increasing concern—among government officials and experts on control systems about potential cyber threats to the control systems that govern our critical infrastructures. As components of control systems increasingly make critical decisions that were once made by humans, the potential effect of a cyber threat becomes more devastating. Such cyber threats could come from numerous sources, ranging from hostile governments and terrorist groups to disgruntled employees and other malicious intruders. Based on interviews and discussions with representatives throughout the electric power industry, the Information Assurance Task Force of the National Security Telecommunications Advisory Committee concluded that an organization with sufficient resources, such as a foreign intelligence service or a well-supported terrorist group, could conduct a structured attack on the electric power grid electronically, with a high degree of anonymity and without having to set foot in the target nation.

In July 2002, the National Infrastructure Protection Center (NIPC) reported that the potential for compound cyber and physical attacks, referred to as "swarming attacks," are an emerging threat to the U.S. critical infrastructure. As NIPC reports, the effects of a swarming attack include slowing or complicating the response to a physical attack. For instance, a cyber attack that disabled the water supply or the electrical system in conjunction with a physical attack could deny emergency services the necessary resources to manage the consequences—such as controlling fires, coordinating actions, and generating light.

Control systems, such as SCADA, can be vulnerable to cyber attacks. Entities or individuals with malicious intent might take one or more of the following actions to successfully attack control systems:

- Disrupt the operation of control systems by delaying or blocking the flow of information through control networks, thereby denying availability of the networks to control system operations.
- Make unauthorized changes to programmed instructions in PLCs, RTUs, or distributed control system controllers, change alarm thresholds, or issue unauthorized commends to control equipment, which could potentially result in damage to equipment (if tolerances are exceeded), premature shutdown of processes (such as prematurely shutting down transmission lines), or even disabling of control equipment.
- Send false information to control system operators, either to disguise unauthorized changes or to initiate inappropriate actions by system operators.
- Modify the control system software, producing unpredictable results.
- Interfere with the operation of safety systems.

In addition, in control systems that cover a wide geographic area, the remote sites are often unstaffed and may not be physically monitored. If such remote systems are physically breached, the attackers could establish a cyber connection to the control network.

Securing Control Systems

Several challenges must be addressed to effectively secure control systems against cyber threats. These challenges include (1) the limitations of current security technologies in securing control systems, (2) the perception that securing control systems may not be economically justifiable, and (3) the conflicting priorities within organizations regarding the security of control systems.

A significant challenge in effectively securing control systems is the lack of specialized security technologies for these systems. The computing resources in control systems that are needed to perform security functions tend to be quite limited, making it very difficult to use security technologies within control system networks without severely hindering performance.

Securing control systems may not be perceived as economically justifiable. Experts and industry representatives have indicated that organizations may be reluctant to spend more money to secure control systems. Hardening the security of control systems would require industries to expend more resources, including acquiring more personnel, providing training for personnel, and potentially prematurely replacing current systems that typically have a lifespan of about 20 years.

Finally, several experts and industry representatives indicate that the responsibility for securing control systems typically includes two separate groups: IT security personnel and control system engineers and operators. IT security personnel tend to focus on securing enterprise systems, while control system engineers and operators tend to be more concerned with the reliable performance of their control systems. Further, they indicate that, as a result, those two groups do not always fully understand each other's requirements and collaborate to implement secure control systems.

STEPS TO IMPROVE SCADA SECURITY

The President's Critical Infrastructure Protection Board and the Department of Energy (DOE) have developed the steps outlined below to help organizations improve the security of their SCADA networks. DOE (2001) points out that these steps are not meant to be prescriptive or all-inclusive. However, they do address essential actions to be taken to improve the protection of SCADA networks. The steps are divided into two categories: specific actions to improve implementation, and actions to establish essential underlying management processes and policies.

21 Steps to Increase SCADA Security

The following steps focus on specific actions to be taken to increase the security of SCADA networks (DOE, 2001).

1. Identify all connections to SCADA networks.

Conduct a thorough risk analysis to assess the risk and necessity of each connection to the SCADA network. Develop a comprehensive understanding of all connections to the SCADA network and of how well those connections are protected. Identify and evaluate the following types of connections:

- Internal local-area and wide-area networks, including business networks
- The Internet
- Wireless network devices, including satellite uplinks
- Modem or dial-up connections
- Connections to business partners, vendors, or regulatory agencies

2. Disconnect unnecessary connections to the SCADA network.

To ensure the highest degree of security of SCADA systems, isolate the SCADA network from other network connections to as great a degree as possible. Any connection to another network introduces security risks, particularly if the connection creates a pathway from or to the Internet. Although direct connections with other networks may allow important information to be passed efficiently and conveniently, insecure connections are simply not worth the risk; isolation of the SCADA network must be a primary goal to provide needed protection. Strategies such as utilization of "demilitarized zones" and data warehousing can facilitate the secure transfer of data from the SCADA network to business networks. However, strategies must be designed and implemented properly to avoid introduction of additional risk through improper configuration.

3. Evaluate and strengthen the security of any remaining connections to the SCADA networks.

Conduct penetration testing or vulnerability analysis of any remaining connections to the SCADA network to evaluate the protection posture associated with these pathways. Use this information in conjunction with risk management processes to develop a robust protection strategy for any pathways to the SCADA network. Since the SCADA network is only as secure as its weakest connecting point, it is essential to implement firewalls, intrusion detection systems (IDSs), and other appropriate security measures at each point of entry. Configure firewall rules to prohibit access from and to the SCADA network, and be as specific as possible when permitting approved connections. For example, an independent system operator should not be granted blanket network access simply because there is a need for a connection to certain components of the SCADA system. Strategically place IDSs at each entry point to alert security personnel of potential breaches of network security. Organization management must understand and accept responsibility or risks associated with any connection to the SCADA network.

4. Harden SCADA networks by removing or disabling unnecessary services

SCADA control servers built on commercial or open-source operating systems can be exposed to attacks by default network services. To the greatest degree possible, remove or disable unused services and network demons to reduce the risk of direct attack. This is particularly important when SCADA networks are interconnected with other networks. Do not permit a service or feature on a SCADA network unless a thorough risk assessment of the consequences of allowing the service/feature shows that the benefits of the service/feature far outweigh the potential for vulnerability exploitation. Examples of services to remove from SCADA networks include automated meter-reading/remote billing systems, email services, and Internet access. An example of a feature to disable is remote maintenance. Numerous secure configurations such as the National Security Agency's series of security guides are available. Additionally, work closely with SCADA vendors to identify secure configurations and coordinate any and all changes to operational systems to ensure that removing or disabling services does not cause downtime, interruption of service, or loss of support.

5. Do not rely on proprietary protocols to protect your system.

Some SCADA systems are unique, proprietary protocols for communications between field devices and servers. Often the security of SCADA systems is based solely on the secrecy of these protocols. Unfortunately, obscure protocols provide very little "real" security. Do not rely on proprietary protocols or factory default configuration settings to protect your system. Additionally, demand that vendors disclose any backdoors or vendor interfaces to your SCADA systems, and expect them to provide systems that are capable of being secured.

6. Implement the security features provided by device and system vendors.

Older SCADA systems (most systems in use) have no security features whatsoever. SCADA system owners must insist that their system vendor implement security features in the form of product patches or upgrades. Some newer SCADA devices are shipped with basic security features, but these are usually disabled to ensure ease of installation.

Analyze each SCADA device to determine whether security features are present. Additionally, factory default security settings (such as in computer network firewalls) are often set to provide maximum usability but minimal security. Set all security features to provide the maximum security only after a thorough risk assessment of the consequences of reducing the security level.

7. Establish strong controls over any medium that is used as a backdoor into the SCADA network.

Where backdoors or vendor connections do exist in SCADA systems, strong authentication must be implemented to ensure secure communications. Modems, and wireless and wired networks used for communications and maintenance, represent a significant vulnerability to the SCADA network and remote sites. Successful "war dialing" or "war driving" attacks could allow an attacker to bypass all other controls and have direct access to the SCADA network or resources. To minimize the risk of such attacks, disable inbound access and replace it with some type of callback system.

8. Implement internal and external intrusion detection systems and establish 24-hour-a-day incident monitoring.

To be able to effectively respond to cyber attacks, establish an intrusion detection strategy that includes alerting network administrators of malicious network activity originating from internal or external sources. Intrusion detection system monitoring is essential 24 hours a day; this capability can be easily set up through a pager. Additionally, incident response procedures must be in place to allow an effective response to any attack. To complement network monitoring, enable logging on all systems and audit system logs daily to detect suspicious activity as soon as possible.

9. Perform technical audits of SCADA devices and networks and any other connected networks to identify security concerns.

Technical audits of SCADA devices and networks are critical to ongoing security effectiveness. Many commercial and open-sourced security tools are available that allow system administrators to conduct audits of their systems/networks to identify active services, patch level, and common vulnerabilities. The use of these tools will not solve systemic problems but will eliminate the paths of least resistance that an attacker could exploit. Analyze identified vulnerabilities to determine their significance, and take corrective actions as appropriate. Track corrective actions and analyze this information to identify trends. Additionally, retest systems after corrective actions have been taken to ensure that vulnerabilities were actually eliminated. Scan nonproduction environments actively to identify and address potential problems.

10. Conduct physical security surveys and assess all remote sites connected to the SCADA network to evaluate their security.

Any location that has a connection to the SCADA network is a target, especially unmanned or unguarded remote sites. Conduct a physical security survey and inventory access points at each facility that has a connection to the SCADA system. Identify and assess any source of information including remote telephone/computer network/fiber-optic cables that could be tapped; radio and microwave links that are exploitable; computer terminals that could be accessed; and wireless local area network access points. Identify and eliminate single points of failure. The security of the site must be adequate to detect or prevent unauthorized access. Do not allow live network access points at remote, unguarded sites simply for convenience.

11. Establish SCADA Red Teams to identify and evaluate possible attack scenarios.

Establish a Red Team (i.e., a team of security experts) to identify potential attack scenarios and evaluate potential system vulnerabilities. Use a variety of people who can provide insight into weaknesses of the overall network, SCADA system, physical systems, and security controls. People who work on the system every day have great insight into the vulnerabilities of your SCADA network and should be consulted when identifying potential attack scenarios and possible consequences. Also, ensure that the risk from a malicious insider is fully evaluated, given that this represents one of the greatest threats to an organization. Feed information resulting from the Red Team evaluation into risk management processes to assess the information and establish appropriate protection strategies.

The remaining steps focus on management actions to establish an effective cyber security program.

12. Clearly define cyber security roles, responsibilities, and authorities for managers, system administrators, and users.

Organization personnel need to understand the specific expectations associated with protecting information technology resources through the definition of clear and logical roles and responsibilities. In addition, key personnel need to be given sufficient authority to carry out their assigned responsibilities. Too often, cyber security is left up to the initiative of the individual responsible for security, which usually leads to inconsistent implementations and ineffective security. Establish a cyber security organizational structure that defines roles and responsibilities and clearly identifies how cyber security issues are escalated and who is notified in an emergency.

13. Document network architecture and identify systems that serve critical functions or contain sensitive information that requires additional levels of protection.

Develop and document robust information security architecture as part of a process to establish an effective protection strategy. It is essential that organizations design their network with security in mind and continue to have a strong understanding of their network architecture throughout its lifecycle. Of particular importance is an in-depth understanding of the functions that the systems perform and the sensitivity of the stored information. Without this understanding, risk cannot be properly assessed and protection strategies may not be sufficient. Documenting the information security architecture and its components is critical to understanding the overall protection strategy and identifying single points of failure.

14. Establish a rigorous, ongoing risk management process.

A thorough understanding of the risks to network computing resources from denial-of-service attacks and the vulnerability of sensitive information is essential to an effective cyber security program. Risk assessments form the technical basis of this understanding and are critical to formulating effective strategies to mitigate vulnerabilities and preserve the integrity of computing resources. Initially, perform a baseline risk analysis based on current threat assessment to use for developing a network protection strategy. Due to rapidly changing technology and the emergence of new threats on a daily basis, an ongoing risk assessment process is also needed so that routine changes can be made to the protection strategy to ensure it remains effective. Fundamental to risk management is identification of residual risk with a network protection strategy in place and acceptance of that risk by management.

15. Establish a network protection strategy based on the principle of defense-in-depth.

A fundamental principle that must be part of any network protection strategy is *defense-in-depth*. Defense-in-depth must be considered early in the design phase of the development process, and must be an integral consideration in all technical decision making associated with the network. Utilize technical and administrative controls to mitigate threats from identified risks to as great a degree as possible at all levels of the network. Single points of failure must be avoided, and cyber security defense must be layered to limit and contain the impact of any security incidents. Additionally, each layer must be protected against other systems at the same layer. For example, to protect against the inside threat, restrict users to access only those resources necessary to perform their job functions.

16. Clearly identify cyber security requirements.

Organizations and companies need structured security programs with mandated requirements to establish expectations and allow personnel to be held accountable. Formalized policies and procedures are typically used to establish and institutionalize a cyber security program. A formal program is essential for establishing a consistent, standards-based approach to cyber security through an organization and eliminates sole dependence on individual initiative. Polices and procedures also inform employees of their specific cyber security responsibilities and the consequences of failing to meet those responsibilities. They also provide guidance regarding actions to be taken during a cyber security incident and promote efficient and effective actions during a time of crisis. As part of identifying cyber security requirements, include user agreements and notification and warning banners. Establish requirements to minimize the threat from malicious insiders; requirements might include background checks and limitation of network privileges.

17. Establish effective configuration management processes.

A fundamental management process needed to maintain a secure network is configuration management. Configuration management needs to cover both hardware configurations and software configurations. Changes to hardware or software can easily introduce vulnerabilities that undermine network security. Processes are required to evaluate and control any change to ensure that the network remains secure. Configuration management begins with well-tested and -documented security baselines for your various systems.

18. Conduct routine self-assessments.

Robust performance evaluation processes are needed to provide organizations with feedback on the effectiveness of cyber security policy and technical implementation. A sign of a mature organization is one that is able to self-identify issues, conduct root cause analyses, and implement effective corrective actions that address individual and systemic problems. Self-assessment processes that are normally part of an effective cyber security program include routine scanning for vulnerabilities, automated auditing of the network, and self-assessments of organizational and individual performance.

19. Establish system backups and disaster recovery plans.

Establish a disaster recovery plan that allows for rapid recovery from any emergency (including a cyber attack). System backups are an essential part of any plan and allow rapid reconstruction of the network. Routinely exercise disaster recovery plans to ensure that they work and that personnel are familiar with them. Make appropriate changes to disaster recovery plans based on lessons learned from exercises.

20. Senior organizational leadership should establish expectations for cyber security performance and hold individuals accountable for their performance.

Effective cyber security performance requires commitment and leadership from senior managers in the organization. It is essential that senior management establish an expectation for strong cyber security and communicate this to their subordinate managers throughout the organization. It is also essential that senior organizational leadership establish a structure for implementation of a cyber security program. This structure will promote consistent implementation and the ability to sustain a strong cyber security program. It is then important for individuals to be held accountable for their performance as it relates to cyber security. This includes managers, system administrators, technicians, and users/operators.

21. Establish policies and conduct training to minimize the likelihood that organizational personnel will inadvertently disclose sensitive information regarding SCADA system design, operations, or security controls.

Release data related to the SCADA network only on a strict, need-to-know basis, and only to persons explicitly authorized to receive such information. "Social engineering," the gathering of information about a computer or computer network via questions to naïve users, is often the first step in a malicious attack on computer networks. The more information revealed about a computer or computer network, the more vulnerable the computer/network is. Never divulge data revealed to a SCADA network, including the names and contact information of system operators/administrators, computer operating systems, and/or physical and logical locations of computers and network systems over telephones or to personnel unless they are explicitly authorized to receive such information. Any requests for information by unknown persons need to be sent to a central network security location for verification and fulfillment. People can be a weak link in an otherwise secure network. Conduct training and information awareness campaigns to ensure that personnel remain diligent in guarding sensitive network information, particularly their passwords.

REFERENCES

DOE. 2001. *21 Steps to Improve Cyber Security of SCADA Networks.* Washington, DC: Department of Energy.

Ezell, B. C. 1998. *Risks of Cyber Attack to Supervisory Control and Data Acquisition.* Charlottesville, VA: University of Virginia.

FBI. 2000. *Threat to Critical Infrastructure.* Washington, DC: Federal Bureau of Investigation.

FBI. 2004. *Ninth Annual Computer Crime and Security Survey.* FBI: Computer Crime Institute and Federal Bureau of Investigation, at www.goesi.com/press.

Gellman, B. 2002. Cyber-attacks by Al Qaeda feared: Terrorists at threshold of using Internet as tool of bloodshed, experts say. *Washington Post,* June 27.

Harris, J. 2005. EPA needs to determine what barriers prevent water systems from securing known SCADA vulnerabilities. Final briefing report at the U.S. EPA. Washington, DC.

NIPC. 2002. *National Infrastructure Protection Center Report.* Washington, DC: National Infrastructure Protection Center.

Stamp, J. et al. 2003. *Common Vulnerabilities in Critical Infrastructure Control Systems.* 2nd ed. Albuquerque, NM: Sandia National Laboratories.

U.S. General Accounting Office. 2003. *Critical Infrastructure Protection: Challenges in Securing Control System.* Washington, DC: U.S. General Accounting Office.

8

Emergency Response

We're in uncharted territory.

—*Rudi Giuliani, September 11, 2001*

When New York Mayor Rudi Giuliani made the above statement to Police Commissioner Bernard Kerik at the World Trade Center site, September 11, 2001, to a degree, one of the first (and not to be forgotten) gross understatements of the twenty-first century had been uttered. Indeed, for citizens of the United States of America, the 9/11 events placed our levels of consciousness, awareness, fear, and questions of what to do next in "uncharted territory." Actually, when you get right down to it, 9/11 generated more questions than anything else. Many are still asking the following questions today:

- Why?
- Why would anyone have the audacity to attack the United States?
- What kind of cold-blooded killers would even think of conducting such an event?
- Who were those Islamic radicals who perpetrated 9/11?
- What were the terrorists' goals?
- Why were we not ready for such an attack?
- Why had we not foreseen such an event?
- Why were our emergency responders so undermanned, ill prepared, and ill equipped to handle such a disaster?
- What took the military fighter planes so long to respond?
- What did our government really know (if anything) before the events occurred?
- Could anyone have prevented it?
- Why us? Why anyone?

These and several other questions continue to resonate today; no doubt they will continue to haunt us for some time to come.

Maybe we ask post-9/11-related questions because of who we are, what we are, and what we are not. That is, because we are Americans we are free, uninhibited thinkers who think what we say and say what we think—isn't America great! Most Americans are soft hearted and sympathetic to those in need—compassion is the very nature and soul of being American. Americans are not born terrorists. They are not born into a terrorist regime, and they are not raised with fear in their hearts—they are not afraid every time they leave their homes and go about their daily business. Suicide bombers and other terrorists are those that occupy some other, faraway place that is definitely not America, and they are definitely not American. Right?

Notwithstanding exceptions to the rule, such as Timothy McVeigh (a so-called red-blooded American, born and raised in America) and the person who mailed the anthrax, terrorism was foreign to us.

Today, from a safety/security point a view, based on the events of 9/11 and the anthrax events, we should no longer be asking why. Instead, we should not waste our time, money, and energy asking why or pointing a finger at our government, military, 9/11 emergency responders, and/or the terrorists. We should stop asking why and start asking what if. In chapter 6, in regards to security preparedness, we pointed to the need to ask what-if questions. It was pointed out, simply, that what-if analysis is a proactive approach used to prevent or mitigate certain disasters and extreme events generated by humans or nature. Obviously, asking and properly answering what-if questions has little effect on preventing natural disasters, such as earthquakes, tornadoes, hurricanes (Katrina-type events), and others. On the other hand, it is true that what-if questions, when properly posed and answered (with results), can reduce the death toll and overall damage caused by these natural disasters. We are certainly aware that these natural events are possible, probable, and likely, and their effects can be horrendous. The irony is apparent, however, because few of us are actually willing to move away from or out of earthquake zones, hurricane and tornado alleys, and floodplains to live somewhere else.

The fact is we do not possess a crystal ball to foretell the future. What-if questions prepare us to react and respond to certain contingencies. And respond we must, because there are certain events we simply can't prevent.

WATER SYSTEM CONTINGENCY PLANNING

Emergency response planning—or contingency planning—for extreme events has long been a standard practice for safety professionals in water/wastewater systems operations. For many years, prudent practices have required consideration of the potential impact of severe natural events, including earthquakes, tornadoes, volcanoes, floods, hurricanes, and blizzards. These possibilities have been included in both water and wastewater infrastructure emergency preparedness and disaster response planning. In addition, many water utilities have considered the potential consequences of workplace violence. Today, as this text has pointed out, there is a new focus of concern: the potential effects of intentional acts by domestic or international terrorists.

As a result, the security paradigm has not necessarily changed, but instead has been adjusted—reasonable, necessary, and sensible accommodations have been and continue to be made. Because we cannot foresee all future intentional acts of terrorism, we must be prepared to shift from the proactive to the reactive mode on short notice—in some cases, on very short notice. Accordingly, we must be prepared to respond to, react to, and mitigate what we can't prevent.

EMERGENCY RESPONSE PLANNING (ERP): STANDARD TEMPLATE

The goals of an Emergency Response Plan (ERP) are to document and understand the steps needed to carry out the following:

- Rapidly restore water/wastewater service after an emergency
- Minimize water/wastewater system damage
- Minimize impact and loss to customers
- Minimize negative impacts on public health and employee safety
- Minimize adverse effects on the environment
- Provide emergency public information concerning customer service
- Provide water/wastewater system information for first responders and other outside agencies

The Environmental Protection Agency (EPA) developed the *Large Water System Emergency Response Outline: Guidance to Assist Community Water Systems in Complying with the Public Health and Bioterrorism Preparedness and Response Act of 2002* (July 2003). This template provides guidance and recommendations to aid facilities in the preparation of emergency response plans under the PL 107-188. The template is provided below. Note that although this template was specifically designed for water systems, it can serve both water and wastewater needs.

Water System ERP Template

I. Introduction

Safe and reliable drinking water is vital to every community. ERP is an essential part of managing a drinking water system. The introduction should identify the requirement to have a documented EFP, identify the goal(s) of the plan (e.g., be able to quickly identify an emergency and initiate timely and effective response action, be able to quickly respond and repair damages to minimize system downtime), and explain how access to the plan is limited. Plans should be numbered for control. Recipients should sign and date a statement that includes their (1) ERP number, (2) agreement not to reproduce the ERP, and (3) signature indicating that they have read the ERP.

An ERP does not necessarily need to be one document. They may consist of an overview document, individual emergency action procedures, checklists, additions to existing operations manuals, appendices, and so on. There may be separate, more detailed plans for specific incidents. There may be plans that do not include particularly sensitive information and those that do. Existing applicable documents should be referenced in the ERP (e.g., chlorine Risk Management Program and contamination response).

II. Emergency Planning Process

Planning Partnerships The planning process should include those parties who will need to help the utility in an emergency situation (e.g., first responders, law enforcement, public health officials, nearby utilities, local emergency planning committees, and testing labs). Partnerships should include entities from the water utility department up through local, state, regional, and federal agencies, as applicable and appropriate, and could also document compliance with governmental requirements.

General Emergency Response Policies, Procedures, Actions, and Documents This section is a short synopsis of the overall emergency management structure, namely, how other utility emergency response, contingency, and risk management plans fit into the ERP for water emergencies. Applicable policies, procedures, actions plans, and reference documents should be cited. Polices should include interconnected agreements with adjacent communities and descriptions of ways the ERP may affect them. Policies should also address how to handle services to other public utility providers such as gas, electric, and so on.

Scenarios Use your vulnerability assessment (VA) findings to identify specific emergency action steps required for response, recovery, and remediation for each of the five incident types (if applicable) outlined in *the Guidance for Water Utility Response, Recovery & Remediation Actions for Man-Made and/or Technological Emergencies.* In this section, a short paragraph referencing the VA and findings should be provided. Specific details identifying vulnerabilities should not be included. In Section V of this plan, specific emergency action procedures addressing each of the incident types should be addressed.

III. Emergency Response Plan: Policies

System-specific Information In an emergency, water systems need to have basic information for system personnel and external parties such as law enforcement, emergency responders, repair contractors/vendors, the media, and others. The information needs to be clearly formatted and readily accessible so system staff can find and distribute it quickly to those who may be involved in responding to the emergency. Basic information that may be presented in the emergency response plan includes the system's ID number, system name, system address or location, directions to the system, population served, number of service connections, system owner, and information about the person in charge of managing the emergency. Distribution maps, detailed plan drawings, site plans, source water locations, and operations manuals may be attached to this plan as appendices or referenced. A list of system-specific information follows.

1. PWS ID, owner, and contact person
2. Population served and service connections
3. System components including (a) pipes and constructed conveyances; (b) physical barriers; (c) isolation valves; (d) water collection, pretreatment, treatment, storage, and distribution facilities; (e) electronic, computer, or other automated systems that are utilized by the public water system, (f) emergency power generators (onsite and portable), (g) the use, storage, or handling of various chemicals; (h) the operation and maintenance of such system components.

Identification of Alternative Water Sources Determine the following:

1. Amount of water needed for various durations
2. Emergency water shipments
3. Emergency water supply sources
4. Identification of alternate storage and treatment sources
5. Regional aid agreements (interconnections)

Also consider in this section a discussion of backup wells, adjacent water systems, certified bulk water haulers, and so on.

Chain-of-command Chart Developed in Coordination with Local Emergency Planning Committee (Internal and/or External Emergency Responders) Include the following:

1. Contact Name
2. Organization and emergency response responsibility
3. Contact information (hardwire, cell phones, faxes, email)
4. State 24-hour emergency communications center information

Communications Procedures: Who, What, When During most emergencies, it will be necessary to quickly notify a variety of parties both internal and external to the water utility. Using the chain-of-command chart and all appropriate personnel from the lists below, indicate who activates the plan, the order in which notification occurs, and the members of the emergency response team. All contact information should be available for routine updating and readily available. The following lists are not intended to be all inclusive, and they should be adapted to your specific needs.

1. Internal Notification Lists:

   Utilities dispatch

   Water source manager

   Water treatment manager

   Water distribution manager

   Facility managers

   Chief water utility engineer

   Director of water utility

   Data (IT) manager

   Wastewater treatment plant

   Other
2. Local Notification Lists

   Head of local government (i.e., mayor, city manager, chairman of board.)

   Public safety officials—fire, local law enforcement (LLE), police, EMS, safety practitioners. If a malevolent act is suspected, LLE should be immediately notified and in turn will notify the FBI, if required. The FBI is the primary agency for investigating sabotage to water systems or terrorist incidents.

   Other government entities—health, schools, parks, finance, electric
3. External Notification Lists

   State public water supply section regulatory agency (or agencies)

   EPA

   State police

   State health department (lab)

   Critical customers (Special considerations for hospitals, federal, state, and county government centers, etc.)

Service/mutual aid

Water Information Sharing and Analysis Center Residential and commercial customers not previously notified

4. Public/Media Notification: When and How to Communicate

   Effective communication is a key element of emergency response, and a media or communications plan is essential to good communications. Be prepared by organizing basic facts about the crisis and your water system. Develop key messages to use with the media that are clear, brief, and accurate. Make sure your messages are carefully planned and have been coordinated with local and state officials. Considerations should be given to establishing protocols for both field and office staff to respectfully defer questions to the utility spokesperson. Be prepared to list geographic boundaries of the affected area (e.g., west of highway a, east of highway b, north of highway c, and south of highway d) to ensure the public clearly understands the system boundaries.

Personnel Safety This section should provide direction as to how operations staff, emergency responders, and the public should respond to a potential toxic release (e.g., chlorine plume release from a water treatment plant or other chemical agents). Possible reactions include facility evacuation, personnel accountability, proper personnel protective equipment as dictated by the Risk Management Program and Process Safety Management Plan, and procedures for the nearby public.

Equipment The ERP should identify equipment that can obviate or significantly lessen the impact of terrorist attacks or other intentional actions on the public health and protect the safety and supply of drinking water provided to communities and individuals. The water utility should maintain an updated inventory of current equipment and repair parts for normal maintenance work.

Because of the potential for extensive or catastrophic damage that could result from a malevolent act, additional equipment sources should be identified for the acquisition and installation of equipment and repair parts in excess of normal usage. This should be based on the results of the specific scenarios and critical assets identified in the vulnerability assessment that could be destroyed. For example, numerous high-pressure pumps, specifically designed for the water utility, could potentially be destroyed. A certain number of long-lead procurement equipment should be inventoried and the vendor information for such unique and critical equipment maintained. In addition, mutual aid agreements with other utilities, and the equipment available under the agreement, should be addressed. Inventories of current equipment, repair parts, and associated vendors should be indicated under Item 29 (Equipment Needs/Maintenance of Equipment) of Section IV (Emergency Action Procedures).

Property Protection A determination should be made as to what water system facilities should be immediately locked down, specific access control procedures implemented, initial security perimeter established, and/or possible secondary malevolent events considered. The initial act may be a diversionary act.

Training, Exercises, and Drills Emergency response training is essential. The purpose of the training program is to inform employees of what is expected of them during an emergency situation. The level of training on an ERP directly affects how well a utility's employees can respond to an emergency. This may take the form of orientation scenarios, tabletop workshops, functional exercises, or others.

Assessment To evaluate the overall ERP's effectiveness and to ensure that procedures and practices developed under the ERP are adequate and are being implemented, the water utility staff should audit the program on a periodic basis.

IV. Emergency Action Procedures (EAPs)

These are detailed procedures used in the event of an operation emergency or malevolent act. EAPs may be applicable across many different emergencies and are common core elements of the overall municipality ERP (e.g., responsibilities, notifications lists, security procedures) and can be referenced. Procedures might include the following:

- Event classification/severity of emergency
- Responsibilities of emergency director
- Responsibilities of incident commander
- Emergency operations center (EOC) activation
- Division internal communications and reporting
- External communications and notifications
- Emergency telephone list (division internal contacts)
- Emergency telephone list (off-site responders, agencies, state 24-hour emergency phone number, and others to be notified)
- Mutual aid agreements
- Contact list of available emergency contractor services/equipment
- Emergency equipment list (including inventory for each facility)
- Security and access control during emergencies
- Facility evacuation and lockdown and personnel accountability
- Treatment and transport of injured personnel (including for chemical/biological exposure)
- Chemical records to compare against historical results for baseline
- List of available labs for emergency use
- Emergency sampling and analysis (chemical/biological, radiological)
- Water use restrictions during emergencies

- Alternate temporary water supplies during emergencies
- Isolation plans for supply, treatment, storage, and distribution systems
- Mitigation plans for neutralizing, flushing, and disinfecting tanks, pump stations, or distribution systems, including shock chlorination
- Protection of vital records during emergencies
- Record keeping and reporting (for requirements determined by the Federal Emergency Management Agency, Occupational Safety and Health Administration, EPA, and others. It is important to maintain accurate financial records of expenses associated with the emergency event for possible federal reimbursement.)
- Emergency program training, drills, and tabletop exercises
- Assessment of emergency management plan and procedures
- Crime scene preservation training and plans
- Communication plans: police, fire, local government, media
- Administration and logistics, including EOC, when established
- Equipment needs/maintenance of equipment
- Recovery and restoration of operations
- Emergency event closeout and recovery

V. Incident-specific EAPs

Incident-specific EAPs are action procedures that identify specific steps in responding to an operational emergency or malevolent act. *The Guidance for Water Utility Response, Recovery & Remediation Actions* identifies three major steps in developing procedures—response, recovery, and remediation—with a list of initial and recovery notifications required. *Response* refers to actions immediately following awareness of the incident, *recovery* refers to actions to bring the system back into operations, and *remediation* refers to long-term restoration actions. When developing an EAP for those incidents identified in Section V.2, the EAP must consider the impact of the incident on system elements and the potential impacts on upstream and downstream components of the incident location. If during the VA process a specific incident type was judged as not credible then it should be noted why it was not applicable to the ERP. If additional incident types were identified, then these should be included in the ERP. For those that use the Sandia National Laboratory methodology (Risk Assessment Methodology—Water) the adversary sequence diagrams provide incident-specific malevolent acts, which may fit under Section V.2.

A. General response to terrorist threats (other than bomb threats and incident-specific threats)
B. Incident-specific response to man-made or technological emergencies
   1. Contamination event (articulated threat with unspecified materials)
   2. Contamination threat at a major event

3. Notification from health officials of potential water contamination
4. Intrusion through supervisory control and data acquisition (SCADA)

C. Significant structural damage resulting from intentional act
D. Customer complaints
E. Severe weather response (snow, ice, temperature, lightning)
F. Flood response
G. Hurricane and/or tornado response
H. Fire response
I. Explosion response
J. Major vehicle accident response
K. Electrical power outage response
L. Water supply interruption response
M. Transportation accident response (barge, plane, train, semi-trailer/tanker)
N. Contaminated/tampered-with water treatment chemicals
O. Earthquakes response
P. Disgruntled employees response (i.e., workplace violence)
Q. Vandals response
R. Bomb threat response
S. Civil disturbance (riot, strike)
T. Armed intruder response
U. Suspicious mail handling and reporting
V. Hazardous chemical spill release response (including Material Safety Data Sheets)
W. Cyber-security/ SCADA system attack response (other than incident specific, e.g., hacker)

VI. Next Steps

A. Plan review and approval
B. Practice and plan to update (as necessary; once every year is recommended) the following:
   1. Training requirements
   2. Persons responsible for conducting training, exercises, and emergency drills
   3. Update and assessment requirements
   4. Incident-specific requirements

VII. Annexes

A. Facility and location information
   1. Facility maps
   2. Facility drawings
   3. Facility descriptions/layout

VIII. References and Links

- Department of Homeland Security, www.dhs/gov/dhspublic
- Environmental Protection Agency, www.epa.gov
- The American Water Works Association, www.awwa.org
- The Center for Disease Control and Prevention, www.bt.cdc.gov
- Federal Emergency Management Agency, ww.fema.gov
- Local Emergency Planning Committees, www.epa.gov/ceppo/epclist.htm

WATER UTILITY RESPONSE, RECOVERY, AND REMEDIATION GUIDELINES

As mentioned, the EPA has been given the responsibility under Presidential Decision Directive 63 for working with the water sector (including water and wastewater utilities) to provide for the protection of the nation's critical water infrastructure, including the systems used to collect, treat, and distribute potable water. The EPA has a similar responsibility for wastewater operations. These critical infrastructures are fundamental to the public health and welfare and are subject to both natural and man-made disasters. Such disasters could place surrounding areas and populations at significant risk. In October 2001, the EPA established an internal Water Protection Task Force to ensure that activities to protect and secure water supply infrastructure are comprehensive and carried out expeditiously. This guidance supports the Task Force's mission of providing information in an expeditious manner to public and private water utilities that can be used to protect public health and critical water infrastructure.

This guidance was developed for the following five different incident types:

- Threat of or actual intentional contamination of the water system
- Threat of contamination at a major event
- Notification from health officials of potential water contamination
- Intrusion through SCADA
- Significant structural damage resulting from an intentional act

While this guidance is oriented toward these five incident types, it should also serve as a guide for response, recovery, and remediation actions for other threatened or actual intentional acts that would affect the safety or security of the water system.

Response Planning

This response, recovery, and remediation guidance to intentional acts can be used to supplement existing water utility emergency operations plans developed to prepare for and respond to natural disasters and emergencies. The EPA (2002) recommends that established policies and procedures contained in existing plans be used to the maximum extent while incorporating the recommendations in this guidance. (See figure 8.1.)

EPA RECOMMENDATIONS

1. Contamination Event: Articulated Threat with Unspecified Material

Event Description. This event is based on the threat of intentional introduction of a contaminant into the water system (at any point within the system) without specification of the contaminant into the water system (at any point within the system) without specification of the contaminant by the perpetrator.

INITIAL NOTIFICATION

- Notify local law enforcement.
- Notify local FBI field office.
- Notify National Response Center.
- Notify local/state emergency management organization.
- Notify ISAC.
- Notify other associated system authorities (wastewater, water).
- Notify local government official.
- Notify local/state health and/or environmental department.
- Notify critical care facilities.
- Notify employees.
- Consider when to notify customers and what notifications to issue.
- Notify governor.

RESPONSE ACTIONS

1. Source water
   - Increase sampling at or near system intakes.
   - Consider whether to isolate the water source if possible.
2. Drinking water treatment facility
   - Preserve latest full battery background test as baseline.
   - Increase sampling efforts.
   - Consider whether to continue normal operations (if determination is made to reduce or stop water treatment, provide notification to customers/issue alerts).
   - Coordinate alternative water supply.
3. Water distribution/storage
   - Consider whether to isolate the water in the affected area if possible.
4. Wastewater collection system
   - Assess what to do with potentially contaminated water within the system based on contaminant, contaminant concentration, potential for system contamination, and ability to bypass treatment plant.
   - If bypassed, notify local and appropriate state authorities and downstream users. Increase monitoring of receiving stream.

FIGURE 8.1
EPA recommendations

5. Wastewater treatment facility
   - Preserve latest full battery background test as baseline.
   - Increase sampling efforts.
   - Consider whether to continue normal operations (if determination is made to reduce or stop water treatment, provide notification to customers/issue alerts).

RECOVERY ACTIONS

Recovery actions should begin once the contaminant is through the system.

RECOVERY NOTIFICATIONS

- Notify customers.
- Notify media.
- Notify ISAC.

APPROPRIATE UTILITY ELEMENTS

- Sample appropriate system elements (storage tanks, filters, sediment basins, solids handling) to determine if residual contamination exists.
- Flush system based on results of sampling.
- Monitor health of employees.
- Plan for appropriate disposition of personal protection equipment (PPE) and other equipment.

REMEDIATION ACTIONS

- Based on sampling results, assess need to remediate storage tanks, filters, sediment basins, solids handling.
- Plan for appropriate disposition of PPE and other equipment.
- If wastewater treatment plant was bypassed, sample and establish monitoring regime for receiving stream and potential remediation based on sampling results.

Note: Response, recovery, and remediation actions may be tailored to a specified (identified) material if the physical properties for the material are known.

II. CONTAMINATION THREAT AT A MAJOR EVENT

Event Description. This event is based on the threat of or actual intentional introduction of a contaminant into the water system at a sports arena, convention center, or similar facility.

Initial Notifications

- Notify local law enforcement.
- Notify local FBI field office.
- Notify National Response Center.
- Notify ISAC.
- Notify local/state emergency management organization.
- Notify wastewater facility.

(Continues)

- Notify governor.
- Notify other associated system authorities (wastewater, water).
- Notify local government official.
- Notify local/state health and/or environmental department.
- Notify critical care facilities.
- Notify employees.
- Consider when to notify customers and what notification to issue.

RESPONSE ACTIONS

1. Source water
   - No recommended action to take
2. Drinking water treatment facility
   - No recommended action to take
3. Water distribution/storage
   - Coordinate isolation of water.
   - Assist in plan for draining the contained water.
   - Assist in developing a plan for sampling water for potential contamination based on threat notification.
   - Provide alternate water source.
4. Wastewater treatment/collection system
   - Coordinate acceptance of isolated water.
   - Monitor accepted water.
   - Assist in plan for draining the contained water.
   - Assist in developing a plan for sampling water for potential contamination based on threat notification.

RECOVERY ACTIONS

Recovery actions should begin once the contaminant is through the system.

RECOVERY NOTIFICATION

- Notify customers in the area of the facility of actions to take.
- Notify customers in affected area once contaminant-free clean water is reestablished.
- Notify downstream users such as water suppliers, irrigators, and electric generating plants.

WATER DISTRIBUTION/STORAGE

- Consider flushing system via hydrants in distribution systems.

REMEDIATION ACTIONS

1. Water distribution/storage
   - Assess need to decontaminate or replace distribution system components.

FIGURE 8.1
(Continued)

2. Wastewater treatment plant
   - On the basis of sampling results, assess need to remediate storage tanks, filters, sediment basins, solids handling.
   - Plan for appropriate disposition of PPE and other equipment.
   - If wastewater treatment plant was bypassed, sample and establish monitoring regime for receiving stream and potential remediation based on sampling results.

III. NOTIFICATION FROM HEALTH OFFICIALS OF POTENTIAL WATER CONTAMINATION

Event Description. This event is based on the water utility being notified by public health officials of potential contamination based on symptoms of patients.

INITIAL NOTIFICATION

- Ask notifying official who else has been notified and request information on symptoms, potential contaminants, and potential area affected.
- Notify local law enforcement.
- Notify local FBI field office.
- Notify National Response Center.
- Notify local/state emergency management organization.
- Notify other associated system authorities (wastewater, water).
- Notify local government official.
- Notify governor.
- Notify local/state health and/or environmental department.
- Notify critical care facilities.
- Notify employees.
- Consider when to notify customers and what notification to issue.
- Notify ISAC.

RESPONSE ACTIONS

1. Source water
   - Increase sampling at or near system intakes.
   - Consider whether to isolate.
2. Drinking water treatment facility
   - Preserve latest full battery background test result as baseline.
   - Increase sampling efforts.
   - Consider whether to continue normal operations (if determination is to reduce or stop water treatment, provide notification to customers/issue alerts).
   - Coordinate alternative water supply (if needed).
3. Water distribution/storage
   - Increase sampling in the area potentially affected and at locations where the contaminant could have migrated to. It is important to consider the time between exposure and onset of symptoms to select sampling sites.

(Continues)

- Consider whether to isolate.
- Consider whether to increase residual disinfectant levels.

4. Wastewater collection system/wastewater treatment facility
    - Increase sampling at pump stations and specifically in the area potentially affected.
    - Assess what to do with potentially contaminated water within the system based on contaminant, contaminant concentration, potential for system contamination, and ability to bypass treatment plant.
    - If bypassed, notify local and appropriate state authorities, downstream users (especially drinking water treatment facilities), and increase monitoring of receiving stream.

RECOVERY ACTIONS

Recovery actions should begin once the contaminant is through the system.

RECOVERY NOTIFICATIONS

- Assist health department with notifications to customers, media, downstream users, and other organizations.

APPROPRIATE UTILITY ELEMENTS

- Sample appropriate system elements (storage tanks, filters, sediment basins, solids handling) to determine if residual contamination exists.
- Flush system based on results of sampling.
- Monitor health of employees.
- Plan for appropriate disposition of personal protection equipment (PPE) and other equipment.

REMEDIATION ACTIONS

- Based on sampling results, assess need to remediate storage tanks, filters, sediment basins, solids handling, and drinking water distribution system.
- Plan for appropriate disposition of PPE and other equipment.
- If wastewater treatment plant was bypassed, sample and establish monitoring regime for receiving stream and potential remediation based on sample results.

Note: Patient symptoms should be used to narrow the list of potential contaminants.

IV. INTRUSION THROUGH SUPERVISORY CONTROL AND DATA ACQUISITION (SCADA)

Event Description. This event is based on internal or external intrusion of the SCADA system to disrupt normal water system operations.

FIGURE 8.1
(Continued)

INITIAL NOTIFICATIONS

- Notify local law enforcement.
- Notify National Infrastructure Protection Center (NIPC) at 1-888-585-9078 (or 202-323-3204/5/6).
- Notify other associated system authorities (wastewater, water).
- Notify employees.
- Notify local FBI field office.
- If the water is assessed to be unfit for consumption, consider when to notify customers and what notification to issue.

RESPONSE ACTIONS

1. Source water
   - Increase sampling at or near system intakes.
   - Consider whether to isolate.
2. Drinking water treatment facility
   - Preserve latest full battery background test as baseline.
   - Increase sampling efforts.
   - Temporarily shut down SCADA system and go to manual operation using established protocol.
   - Consider whether to shut down system and provide alternate water.
3. Water distribution/storage
   - Monitor unmanned components (storage tanks and pumping stations).
   - Consider whether to isolate.
4. Wastewater collection system/treatment facility
   - Temporarily shut down SCADA system and go to manual operation using established protocol.
   - Monitor unmanned components (pumping stations)—required only if wastewater SCADA system is compromised.
   - If SCADA intrusion caused release of improperly treated water, consider whether to continue normal operations (if determination is made to reduce or stop water treatment, provide notification to customers/issue alerts).

RECOVERY ACTIONS

Recovery actions should begin once the intrusion has been eliminated and the contaminant/unsafe water (if this occurs) is through the system.

RECOVERY NOTIFICATIONS

- Notify employees.
- Notify local law enforcement.
- Notify customers and media if the event resulted in contamination and the full range of standard notifications were made.

(Continues)

APPROPRIATE UTILITY ELEMENTS

- With FBI assistance, make an image copy of all system logs to preserve evidence.
- With FBI assistance, check for implanted backdoors and other malicious code and eliminate them before restarting SCADA system.
- Install safeguards before restarting SCADA.
- Bring SCADA system up and monitor system.

REMEDIATION ACTIONS

- Assess/implement additional protections for SCADA system.
- Check for an NIPC water sector warning based on the intrusion that may contain additional protective actions to be considered. NIPC warnings can be found at www.NIPC.gov or at http://www.infrgard.org for secure access Infragard members.

V. SIGNIFICANT STRUCTURAL DAMAGE RESULTING FROM AN INTENTIONAL ACT

Event Description. This event is based on intentional structural damage to water system components to disrupt normal system operations.

INITIAL NOTIFICATIONS

- Notify local law enforcement.
- Notify local FBI field office.
- Notify National Response Center.
- Notify local/state emergency management organization.
- Notify governor.
- Notify ISAC.
- Notify other associated system authorities (wastewater, water).
- Notify local government officials.
- Notify local/state health and/or environmental department.
- Notify critical care facilities.
- Notify employees.
- Consider when to notify customers and what notification to issue.

RESPONSE ACTIONS

1. Source water
   - Deploy damage assessment teams; if damage appears to be intentional, then treat it as a crime scene and consult local/state law enforcement and FBI on evidence preservation.
   - Inform law enforcement and FBI of potential hazardous materials.
   - Coordinate alternative water supply, as needed.

FIGURE 8.1
(Continued)

- Consider increasing security measures.
- Based on extent of damage, consider alternate (interim) treatment schemes to maintain at least some level of treatment.

2. Drinking water treatment system
   - Deploy damage assessment teams; if damage appears to be intentional, then treat it as a crime scene and consult local/state law enforcement and FBI on evidence preservation.
   - Inform law enforcement and FBI of potential hazardous materials.
   - Coordinate alternative water supply, as needed.
   - Consider increasing security measures.
   - Based on extent of damage, consider alternate (interim) treatment schemes to maintain at least some level of treatment.
3. Water distribution/storage
   - Deploy damage assessment team; if damage appears to be intentional, then treat it as a crime scene and consult local/state law enforcement and FBI on evidence preservation.
   - Inform law enforcement and FBI of potential hazardous materials.
   - Coordinate alternative water supply, as needed.
   - Consider increasing security measures.
   - Based on extent of damage, consider alternate (interim) treatment schemes to maintain at least some level of treatment.
4. Wastewater collection system/treatment facility
   - Deploy damage assessment teams; if damage appears to be intentional, then treat it as a crime scene and consult local/state law enforcement and FBI on evidence preservation.
   - Inform law enforcement and FBI of potential hazardous materials.
   - Coordinate alternative water supply, as needed.
   - Consider increasing security measures.
   - Based on extent of damage, consider alternate (interim) treatment schemes to maintain at least some level of treatment.

RECOVERY ACTIONS

- Recovery actions should begin as soon as practical after damaged facility is isolated from the rest of the utility facilities.

RECOVERY NOTIFICATIONS

- Notify employees.
- Notify local law enforcement.
- Notify local FBI office.

Appropriate Utility Elements

- Dependent on the feedback from damage assessment teams
- Implement damage recovery plan

(Continues)

REMEDIATION ACTIONS

- Repair damage.
- Assess need for additional protection/security measures for damaged facility and other critical facilities within the utility.

FIGURE 8.1
(Continued)

REFERENCES

The Guidance for Water Utility Response, Recovery & Remediation Actions for Man-Made and/or Technological Emergencies. 2002. Washington, DC: U.S. EPA Office of Water (4610M) EPA 810-R-02-001. www.epa.gov/safewater.

Large Water System Emergency Response Plan Outline: Guidance to Assist Community Water Systems in Complying with the Public Health Security and Bioterrorism Preparedness and Response Act. 2002. Washington, DC: U.S. EPA 810-F-03-007. www.epa.gov/safewater/security.

9

Security Techniques and Hardware

If you think we are free today, you know nothing about tyranny and even less about freedom.

—*Tom Braun*

Ideally, in a perfect world, water and wastewater infrastructure would be secured in a layered fashion (a.k.a., the barrier approach). Layered security systems are vital. Using the "protection in depth" principle, that is, requiring that an adversary defeat several protective barriers or security layers to accomplish its goal, water/wastewater infrastructure can be made more secure. Protection in depth is a term commonly used by the military to describe security measures that reinforce one another, masking the defense mechanisms from view of intruders, and allowing the defender time to respond to intrusion or attack.

A prime example of the use of the multibarrier approach to ensure security and safety is demonstrated by the practices of the bottled water industry. In the aftermath of 9/11 and the increased emphasis on homeland security, a shifted paradigm of national security and vulnerability awareness has emerged. Recall that in the immediate aftermath of the 9/11 tragedies, emergency responders and others responded quickly and worked to exhaustion. In addition to the emergency responders, bottled water companies responded immediately by donating several million bottles of water to the crews at the crash sites in New York, at the Pentagon, and in Pennsylvania. International Bottled Water Association (IBWA, 2004) reports that "within hours of the first attack, bottled water was delivered where it mattered most; to emergency personnel on the scene who required ample water to stay hydrated as they worked to rescue victims and clean up debris" (p. 2).

Bottled water companies continued to provide bottled water to responders and rescuers at the 9/11 sites throughout the post-event(s) process(es). These patriotic actions by the bottled water companies, however, beg the question, How do we ensure the safety and security of the bottled water provided to anyone? IBWA (2004) has the answer: Using a multibarrier approach, along with other principles, will enhance the safety and security of bottled water. IBWA (2004) describes its multibarrier approach as follows:

> A multi-barrier approach—Bottled water products are produced utilizing a multi-barrier approach, from source to finished product, that helps prevent possible harmful contaminants (physical, chemical or microbiological) from adulterating the finished product as well as storage, production, and transportation equipment. Measures in a multi-barrier approach may include source protection, source monitoring, reverse osmosis, distillation, filtration, ozonation or ultraviolet (UV) light. Many of the steps in a multi-barrier system may be effective in safeguarding bottled water from microbiological and other contamination. Piping in and out of plants, as well as storage silos and water tankers are also protected and maintained through sanitation procedures. In addition, bottled water products are bottled in a controlled, sanitary environment to prevent contamination during the filling operation. (p. 3)

In water/wastewater infrastructure security, protection in depth is used to describe a layered security approach. A protection in depth strategy uses several forms of security techniques and/or devices against an intruder and does not rely on one single defensive mechanism to protect infrastructure. By implementing multiple layers of security, a hole or flaw in one layer is covered by the other layers. An intruder will have to intrude through each layer without being detected in the process. The layered approach implies that no matter how an intruder attempts to accomplish his goal, he will encounter effective elements of the physical protection system.

For example, as depicted in figure 9.1, an effective security layering approach requires that an adversary penetrate multiple, separate barriers to gain entry to a critical target at a water/wastewater facility. As shown in figure 9.1, protection in depth (multiple layers of security) helps to ensure that the security system remains effective in the event of a failure or an intruder bypassing a single layer of security.

Again, as shown in figure 9.1, layered security starts with the outer perimeter (the fence—the first line of physical security) of the facility and goes inward to the facility, the buildings, structures, other individual assets, and finally to the contents of those buildings—the targets.

The area between the outer perimeter and structures or buildings is known as the site. This open site area provides an incomparable opportunity for early identification of an unauthorized intruder and initiation of early warning/response. This open space area is commonly used to calculate the standoff distance; that is, it is the distance between the outside perimeter (public areas to the fence) and the targets or critical assets

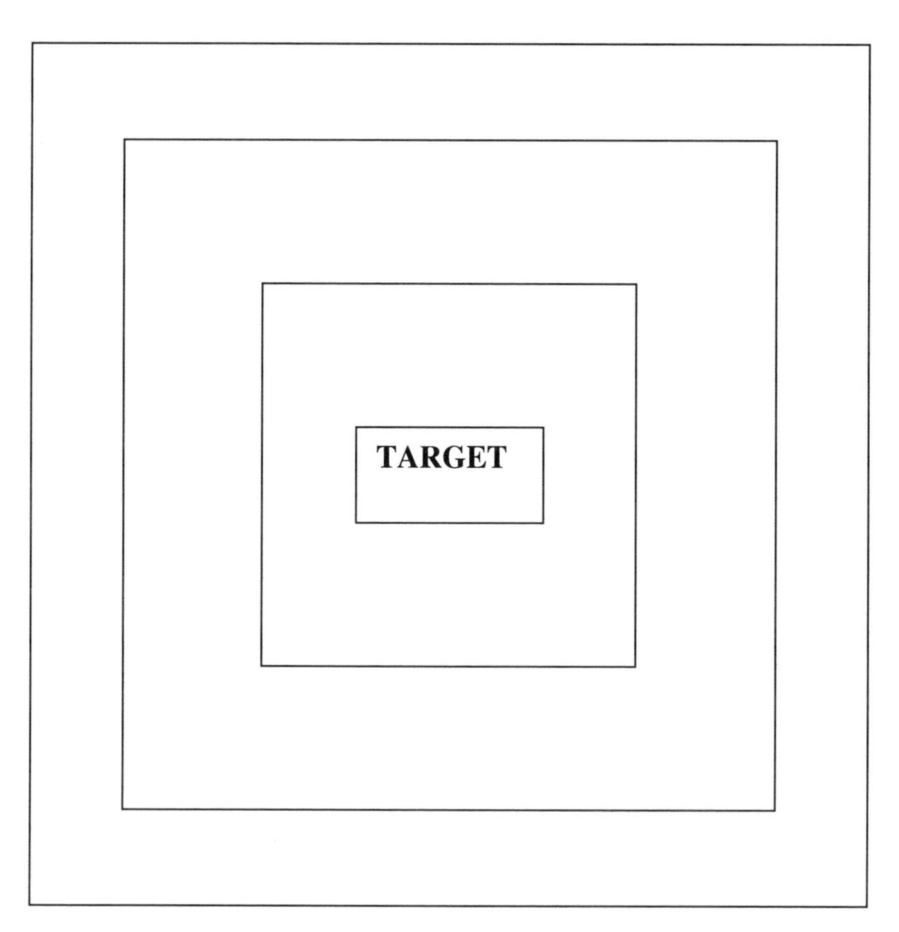

FIGURE 9.1
Layered approach to security

(buildings/structures) inside the perimeter (inside the fence line—the restricted access area).

The open area, between perimeter fence and target (e.g., operations center), if properly outfitted with various security devices, can also provide layered protection against intruders. For example, lighting is a deterrent. Based on personal experience, keeping an open area within the plant site as well lighted at night as would be expected during daylight hours is the rule of thumb. In addition, strategically placed motion detectors along with crash barriers at perimeter gate openings and in front of vital structures are also recommended. Armed, mobile guards who roam the interior of the plant site on a regular basis provide the ultimate in site area security.

The next layer of physical security is the outside wall of the target structure(s) itself. Notwithstanding door, window, and/or skylight entry, walls prevent most intruders from easy entry. If doors can only be entered using card reader access, security is shored up or enhanced to an extent. The same can be said for windows and skylights that are fashioned small enough to prohibit normal human entry. These same "weak" spots in buildings can be bastioned with break-proof or reinforced security glass.

The final layer of security is provided by properly designed interior features of buildings. Examples of these types of features include internal doors and walls, equipment cages, and backup or redundant equipment.

In the discussion above, conditions described referred to perfect world conditions; that is, to those conditions that we would want (i.e., the security manager's proverbial wish list) to be incorporated into the design and installation of new water/wastewater infrastructure. Post-9/11, in a not so perfect world, however, many of the peripheral (fence line) measures described above are more difficult to incorporate into water/wastewater sites/infrastructure. This is not to say that water/wastewater facilities do not have fence lines or fences; most of them do. These fences are designed to keep vandals, thieves, and trespassers out. The problem is that many of these facilities were constructed several years ago, before urban encroachment literally encircled the sites, allowing, at present, little room for security stand backs or setbacks to be incorporated into plants or critical equipment locations. Based on personal observation, many of these fences face busy city streets or closely abut structures outside the fence line. The point is that when one sits down to plan a security upgrade, these factors must be taken into account.

For existing facilities, security upgrades should be based on the results generated from the vulnerability assessment, which characterizes and prioritizes those assets that may be targeted. Those vulnerabilities identified must be protected.

In the following sections, various security hardware and/or devices are described. These devices serve the main purpose of providing security against physical and/or digital intrusion. That is, they are designed to delay and deny intrusion and are normally coupled with detection and assessment technology. However, as mentioned previously, no matter the type of security device or system employed, water/wastewater systems cannot be made immune to all possible intrusions. Simply, when it comes to making anything absolutely secure from intrusion or attack, there is inherently, or otherwise, no silver bullet.

SECURITY HARDWARE/DEVICES

The EPA (2005) groups the water/wastewater infrastructure security devices or products described below into four general categories:

- Physical asset monitoring and control devices
- Cyber protection devices
- Communication/integration devices
- Water monitoring devices

PHYSICAL ASSET MONITORING AND CONTROL DEVICES

Aboveground Outdoor Equipment Enclosures

Water and wastewater systems consist of multiple components spread over a wide area and typically include a centralized treatment plant, as well as distribution or collection system components that are distributed at multiple locations throughout the community. However, in recent years, distribution and collection system designers have favored placing critical equipment—especially assets that require regular use and maintenance—aboveground.

One of the primary reasons for doing so is that locating this equipment aboveground eliminates the safety risks associated with confined space entry, which is often required for the maintenance of equipment located belowground. In addition, space restrictions often limit the amount of equipment that can be located inside, and there are concerns that some types of equipment (such as backflow prevention devices) can, under certain circumstances, discharge water that could flood pits, vaults, or equipment rooms. Therefore, many pieces of critical equipment are located outdoors and aboveground.

Many different system components can be installed outdoors and aboveground. Examples of these types of components could include the following:

- Backflow prevention devices
- Air release and control valves
- Pressure vacuum breakers
- Pumps and motors
- Chemical storage and feed equipment
- Meters
- Sampling equipment
- Instrumentation

Much of this equipment is installed in remote locations and/or in areas where the public can access it.

One of the most effective security measures for protecting aboveground equipment is to place it inside a building. When/where this is not possible, enclosing the equipment or parts of the equipment using some sort of commercial or homemade add-on structure may help to prevent tampering with the equipment. These types of add-on structures or enclosures, which are designed to protect the equipment both from the elements and from unauthorized access or tampering, typically consist of a boxlike structure that is placed over the entire component, or over critical parts of the component (e.g., valves), and is then secured to delay or prevent intruders from tampering with the equipment. The enclosures are typically locked or otherwise anchored to a solid foundation, which makes it difficult for unauthorized personnel to remove the enclosure and access the equipment.

Standardized aboveground enclosures are available in a wide variety of materials, sizes, and configurations. Many options and security features are also available for each type of enclosure, and this allows system operators the flexibility to customize an enclosure for a specific application and/or price range. In addition, most manufacturers can custom-design enclosures if standard, off-the-shelf enclosures do not meet a user's needs.

Many of these enclosures are designed to meet certain standards. For example, the American Society of Sanitary Engineers (ASSE) has developed Standard #1060, Performance Requirements for Outdoor Enclosures for Backflow Prevention Assemblies. If an enclosure will be used to house a backflow preventer, this standard specifies the acceptable construction materials for the enclosure, as well as the performance requirements that the enclosure should meet, including specifications for freeze protection, drainage, air inlets, access for maintenance, and hinge requirements. ASSE #1060 also states that the enclosure should be lockable to enhance security.

Equipment enclosures can generally be categorized into one of the following four main configurations:

- One-piece drop-over enclosures
- Hinged or removable top enclosures
- Sectional enclosures
- Shelters with access locks

All enclosures, including those with integral floors, must be secured to a foundation to prevent them from being moved or removed. Unanchored or poorly anchored enclosures may be blown off the equipment being protected or may be defeated by intruders. In either case, this may result in the equipment beneath the enclosure becoming exposed and damaged. Therefore, ensuring that the enclosure is securely anchored will increase the security of the protected equipment.

The three basic types of foundations that can be used to anchor the aboveground equipment enclosure are concrete footers, concrete slabs-on-grade, or manufactured fiberglass pads. The most common types of foundations utilized for equipment enclosures are standard or slab-on-grade footers; however, local climate and soil conditions may dictate whether either of these types of foundations can be used. These foundations can be either precast or poured in place at the installation site. Once the foundation is installed and properly cured, the equipment enclosure is bolted or anchored to the foundation to secure it in place.

An alternative foundation, specifically for use with smaller hot box enclosures, is a manufactured fiberglass pad known as the Glass Pad. The Glass Pad has the center cut out so that it can be dropped directly over the piece of equipment being enclosed. Once the pad is set level on the ground, it is backfilled over a two-inch flange located around its base. The enclosure is then placed on top of the foundation and is locked in place with either a staple- or a slotted-anchor, depending on the enclosure configuration.

One of the primary attributes of a security enclosure is its strength and resistance to breaking and penetration. Accordingly, the materials from which the enclosure is constructed will be important in determining the strength of the enclosure and thus its usefulness for security applications. Enclosures are typically manufactured of either fiberglass or aluminum. With the exception of the one-piece drop-over enclosure, which is typically fabricated from fiberglass, each configuration described above can be constructed from either material. In addition, enclosures can be custom manufactured from polyurethane, galvanized steel, or stainless steel. Galvanized or stainless steel is often offered as an exterior layer, or "skin," for an aluminum enclosure. Although they are typically utilized in underground applications, precast concrete structures can also be used as aboveground equipment enclosures. However, precast structures are much heavier and more difficult to maneuver than their fiberglass and aluminum counterparts. Concrete is also brittle, and that can be a security concern; however, products (e.g., epoxy coating) can be applied to concrete structures to add strength and minimize security risks. Because precast concrete structures can be purchased from any concrete producers, this document does not identify specific vendors for these types of products.

In addition to the construction materials, enclosure walls can be configured or reinforced to give them added strength. Adding insulation is one option that can strengthen the structural characteristics of an enclosure; however, some manufacturers offer additional features to add strength to exterior walls. For example, while most enclosures are fabricated with a flat wall construction, some vendors manufacture fiberglass shelters with ribbed exterior walls. These ribs increase the structural integrity of the wall and allow the fabrication of standard shelters up to 20 feet in length. Another vendor has developed a proprietary process that uses a series of integrated fiberglass beams that are placed throughout a foam inner core to tie together the interior and exterior walls and roof. Yet another vendor constructs aluminum enclosures with horizontal and vertical redwood beams for structural support.

Other security features that can be implemented on aboveground outdoor equipment enclosures include locks, mounting brackets, tamper-resistant doors, and exterior lighting.

Active Security Barriers (Crash Barriers)

Active security barriers (also known as crash barriers) are large structures that are placed in roadways at entrance and exit points of protected facilities to control vehicle access. These barriers are placed perpendicular to traffic to block the roadway so that the barrier must be moved for traffic to pass. These types of barriers are typically constructed from sturdy materials, such as concrete or steel, such that vehicles cannot penetrate them. They are also designed at a certain height so that vehicles cannot go over them.

The key difference between active security barriers—which include wedges, crash beams, gates, retractable bollards, and portable barricades—and passive security barriers—which include nonmoveable bollards, jersey barriers, and planters—is that active security barriers are designed so that they can be raised and lowered or moved out of the roadway easily to allow authorized vehicles to pass. Many of these types of barriers are designed so that they can be opened and closed automatically (e.g., mechanized gates, hydraulic wedge barriers), while others are easy to open and close manually (e.g., swing crash beams, manual gates). In contrast to active barriers, passive barriers are permanent, nonmovable barriers, and thus they are typically used to protect the perimeter of a protected facility, such as sidewalks and other areas that do not require vehicular traffic to pass them. Several of the major types of active security barriers, such as wedge barriers, crash beams, gates, bollards, and portable/removable barricades, are described below.

Wedge barriers are plated, rectangular steel buttresses approximately two to three feet high that can be raised and lowered from the roadway. When they are in the open position, they are flush with the roadway, and vehicles can pass over them. However, when they are in the closed (armed) position, they project up from the road at a 45-degree angle, with the upper end pointing toward the oncoming vehicle and the base of the barrier away from the vehicle. Generally, wedge barriers are constructed from heavy gauge steel or concrete that contains an impact-dampening iron rebar core that is strong and resistant to breaking or cracking, thereby allowing them to withstand the impact from a vehicle attempting to crash through them. In addition, both of these materials help to transfer the energy of the impact over the barrier's entire volume, thus helping to prevent the barrier from being sheared off its base. Finally, the angle of the barrier impedes any vehicles attempting to drive over it.

Wedge barriers can be fixed or portable. Fixed wedge barriers can be mounted on the surface of the roadway (surface-mounted wedges) or in a shallow mount in the road's surface, or they can be installed completely below the road surface. Surface-mounted wedge barricades operate by rising from a flat position on the surface of the roadway, while shallow-mount wedge barriers rise from their resting position just below the road surface. In contrast, below-surface wedge barriers operate by rising from beneath the road surface. Both the shallow-mounted and surface-mounted barriers require little or no excavation, and thus do not interfere with buried utilities. All three barrier types project above the road surface and block traffic when they are raised into the armed position. Once they are disarmed and lowered, they are flush with the road, thereby allowing traffic to pass. Portable wedge barriers are moved into place on wheels that are removed after the barrier has been set into place.

Installing rising wedge barriers requires preparation of the road surface. Installing surface-mounted wedges does not require that the road be excavated; however, the road surface must be intact and strong enough to allow the bolts anchoring the wedge to the road surface to attach properly. Shallow-mounted and below-surface wedge barricades require excavation of a pit that is large enough to accommodate the wedge structure as well as any arming/disarming mechanisms. Generally, the bottom of the excavation pit is lined with gravel to allow for drainage. Areas not sheltered from rain or surface runoff can install a gravity drain or self-priming pump. Table 9.1 lists the pros and cons of wedge barriers.

Table 9.1. Pros and Cons of Wedge Barriers

Pros	*Cons*
Can be surface mounted or completely installed below the roadway surface	Installations below the surface of the roadway will require construction that may interfere with buried utilities.
Wedge barriers have a quick response time (normally 3.5–10.5 seconds, but it can be 1–3 seconds in emergency situations). Because emergency activation of the barrier causes more wear and tear on the system than normal activation, it is recommended for use only in true emergency situations.	Regular maintenance is needed to keep wedge barrier fully operational.
Surface or shallow-mount wedge barricades can be utilized in locations with a high water table and/or corrosive soils.	Improper use of the system may result in authorized vehicles being hung up by the barrier and damaged. Guards must be trained to use the system properly to ensure that this does not happen. Safety technologies may also be installed to reduce the risk of the wedge activating under an authorized vehicle.
All three wedge barrier designs have a high crash rating, thereby allowing them to be employed for higher security applications. These types of barriers are extremely visible, which may deter potential intruders	

Source: Water and Wastewater Security Product Guide, U.S. EPA, 2005.

Crash beam barriers consist of aluminum beams that can be opened or closed across the roadway. While there are several different crash beam designs, every crash beam system consists of an aluminum beam that is supported on each side by a solid footing or buttress, which is typically constructed from concrete, steel, or some other strong material. Beams typically contain an interior steel cable (usually at least one inch in diameter) to give the beam added strength and rigidity. The beam is connected by a heavy duty hinge or other mechanism to one of the footings so that it can swing or rotate out of the roadway when it is open, and can swing back across the road when it is in the closed (armed) position, blocking the road and inhibiting access by unauthorized vehicles. The nonhinged end of the beam can be locked into its footing, thus providing anchoring for the beam on both sides of the road and increasing the beam's resistance to any vehicles attempting to penetrate it. In addition, if the crash beam is hit by a vehicle, the aluminum beam transfers the impact energy to the interior cable, which in turn transfers the impact energy through the footings and into their foundation, thereby minimizing the chance that the impact will snap the beam and allow the intruding vehicle to pass through.

Crash beam barriers can employ drop-arm, cantilever, or swing beam designs. Drop-arm crash beams operate by raising and lowering the beam vertically across the road. Cantilever crash beams are projecting structures that are opened and closed by extending the beam from the hinge buttress to the receiving buttress located on the opposite side of the road. In the swing beam design, the beam is hinged to the buttress such that it swings horizontally across the road. Generally, swing beam and cantilever designs are used at locations where a vertical lift beam is impractical. For example, the swing beam or cantilever designs are utilized at entrances and exits with overhangs, trees, or buildings that would physically block the operation of the drop-arm beam design.

Installing any of these crash beam barriers involves the excavation of a pit approximately 48 inches deep for both the hinge and the receiver footings. Due to the depth of excavation, the site should be inspected for underground utilities before digging begins. Table 9.2 lists the pros and cons of crash beams.

In contrast to wedge barriers and crash beams, which are typically installed separately from a fence line, gates are often-integrated units of a perimeter fence or wall around a facility. Gates are basically movable pieces of fencing that can be opened and closed across a road. When the gate is in the closed (armed) position, the leaves of the gate lock into steel buttresses that are embedded in a concrete foundation located on both sides of the roadway, thereby blocking access to the roadway. Generally, gate barricades are constructed from a combination of heavy gauge steel and aluminum that can absorb an impact from vehicles attempting to ram through them. Any remaining impact energy not absorbed by the gate material is transferred to the steel buttresses and their concrete foundation.

Table 9.2. Pros and Cons of Crash Beams

Pros	*Cons*
Requires little maintenance, while providing long-term durability	Crash beams have a slower response time (normally 9.5–15.3 seconds, but it can be reduced to 7–10 seconds in emergency situations) than other types of active security barriers, such as wedge barriers. Because emergency activation of the barrier causes more wear and tear on the system than normal activation, it is recommended for use only in true emergency situations.
No excavation is required in the roadway itself to install crash beams.	All three crash beam designs possess a low crash rating relative to other types of barriers, such as wedge barriers, and thus they typically are used for lower security applications.
	Certain crash barriers may not be visible to oncoming traffic and therefore may require additional lighting and/or other warning markings to reduce the potential for traffic to accidentally run into the beam.

Source: Water and Wastewater Security Product Guide, U.S. EPA, 2005.

Gates can utilize a cantilever, linear, or swing design. Cantilever gates are projecting structures that operate by extending the gate from the hinge footing across the roadway to the receiver footing. A linear gate is designed to slide across the road on tracks via a rack and pinion drive mechanism. Swing gates are hinged so that they can swing horizontally across the road.

Installation of the cantilever, linear, or swing gate designs described above involve the excavation of a pit approximately 48 inches deep for both the hinge and receiver footings to which the gates are attached. Due to the depth of excavation, the site should be inspected for underground utilities before digging begins. Table 9.3 lists the pros and cons of gates.

Bollards are vertical barriers at least three feet tall and one to two feet in diameter that are typically set four to five feet apart from each other to block vehicles from passing between them. Bollards can be either fixed in place, removable, or retractable. Fixed and removable bollards are passive barriers that are typically used along building perimeters or on sidewalks to block vehicles while allowing pedestrians to pass. In contrast to passive bollards, retractable bollards are active security barriers that can easily be raised and lowered to allow vehicles to pass between them. Thus, they can be used in driveways or on roads to control vehicular access. When the bollards are raised, they project above the road surface and block the roadway; when they are lowered, they sit flush with the road surface, and thus allow traffic to pass over them. Retractable

Table 9.3. Pros and Cons of Gates

Pros	*Cons*
All three gate designs possess an intermediate crash rating, thereby allowing them to be utilized for medium to higher security applications.	Gates have a slower response time (normally 10–15 seconds, but it can be reduced to 7–10 seconds in emergency situations) than other types of active security barriers, such as wedge barriers. Because emergency activation of the barrier causes more wear and tear on the system than normal activation, it is recommended for use only in true emergency situations.
Requires very little maintenance.	
Can be tailored to blend in with perimeter fencing.	
Gate construction requires no roadway excavation.	
Cantilever gates are useful for roads with high crowns or drainage gutters.	
These types of barriers are extremely visible, which may deter intruders.	
Gates can also be used to control pedestrian traffic.	

Source: Water and Wastewater Security Product Guide, U.S. EPA, 2005.

bollards are typically constructed from steel or other materials that have a low weight-to-volume ratio so that they require low power to raise and lower. Steel is also more resistant to breaking than a more brittle material, such as concrete, and is better able to withstand direct vehicular impact without breaking apart.

Retractable bollards are installed in a trench dug across a roadway—typically at an entrance or gate. Installing retractable bollards requires preparing the road surface. Depending on the vendor, bollards can be installed either in a continuous slab of concrete or in individual excavations with concrete poured in place. The required excavation for a bollard is typically slightly wider and slightly deeper than the bollard height when extended aboveground. The bottom of the excavation is typically lined with gravel to allow drainage. The bollards are then connected to a control panel, which controls the raising and lowering of the bollards. Installation typically requires mechanical, electrical, and concrete work; if utility personnel with these skills are available, then the utility can install the bollards themselves. Table 9.4 lists the pros and cons of retractable bollards.

Portable/removable barriers, which can include removable crash beams and wedge barriers, are mobile obstacles that can be moved in and out of position on a roadway. For example, a crash beam may be completely removed and stored off-site when it is not needed. An additional example would be wedge barriers equipped with wheels that can be removed after the barricade is towed into place.

When portable barricades are needed, they can be moved into position rapidly. To provide them with added strength and stability, they are typically anchored to buttress boxes that are located on either side of the road. These buttress boxes, which may or

Table 9.4. Pros and Cons of Retractable Bollards

Pros	*Cons*
Bollards have a quick response time (normally 310 seconds, but it can be reduced to 1–3 seconds in emergency situations).	Bollard installations will require construction below the surface of the roadway, which may interfere with buried utilities.
Bollards have an intermediate crash rating, which allows them to be utilized for medium to higher security applications.	Some maintenance is needed to ensure barrier is free to move up and down.
	The distance between bollards must be decreased (i.e., more bollards must be installed along the same perimeter) to make these systems effective against small vehicles (i.e., motorcycles).

Source: Water and Wastewater Security Product Guide, U.S. EPA, 2005.

may not be permanent, are usually filled with sand, water, cement, gravel, or concrete to make them heavy and aid in stabilizing the portable barrier. In addition, these buttresses can help dissipate any impact energy from vehicles crashing into the barrier itself.

Because these barriers are not anchored into the roadway, they do not require excavation or other related construction for installation. In contrast, they can be assembled and made operational in a short period of time. The primary shortcoming to this type of design is that these barriers may move if they are hit by vehicles. Therefore, it is important to carefully assess the placement and anchoring of these types of barriers to ensure that they can withstand the types of impacts that may be anticipated at that location. Table 9.5 lists the pros and cons of portable/removable barricades.

Because the primary threat to active security barriers is that vehicles will attempt to crash through them, their most important attributes are their size, strength, and crash

Table 9.5. Pros and Cons of Portable/Removable Barricades

Pros	*Cons*
Installing portable barricades requires no foundation or roadway excavation	Portable barriers may move slightly when hit by a vehicle, resulting in a lower crash resistance.
Can be moved in and out of position in a short period of time.	Portable barricades typically require 7.75–16.25 seconds to move into place, and thus they are considered to have a medium response time when compared with other active barriers.
Wedge barriers equipped with wheels can be easily towed into place.	
Minimal maintenance is needed to keep barriers fully operational.	

Source: Water and Wastewater Security Product Guide, U.S. EPA, 2005.

resistance. Other important features for an active security barrier are the mechanisms by which the barrier is raised and lowered to allow authorized vehicle entry, weather resistance, and safety features.

Alarms

An alarm system is a type of electronic monitoring system that is used to detect and respond to specific types of events—such as unauthorized access to an asset or a possible fire. In water and wastewater systems, alarms are also used to alert operators when process operating or monitoring conditions go out of preset parameters (i.e., process alarms). These types of alarms are primarily integrated with process monitoring and reporting systems (e.g., supervisory control and data acquisition [SCADA] systems). Note that this discussion does not focus on alarm systems that are not related to a utility's processes.

Alarm systems can be integrated with fire detection systems, intrusion detection systems (IDSs), access control systems, or closed-circuit television (CCTV) systems, such that these systems automatically respond when the alarm is triggered. For example, a smoke detector alarm can be set up to automatically notify the fire department when smoke is detected, or an intrusion alarm can automatically trigger cameras to turn on in a remote location so that personnel can monitor that location.

An alarm system consists of sensors that detect different types of events; an arming station that is used to turn the system on and off; a control panel that receives information, processes it, and transmits the alarm; and an annunciator that generates a visual and/or audible response to the alarm. When a sensor is tripped, it sends a signal to a control panel, which triggers a visual or audible alarm and/or notifies a central monitoring station. A more complete description of each of the components of an alarm system is provided below.

Detection devices (also called sensors), are designed to detect a specific type of event (such as smoke, intrusion, etc.). Depending on the type of event they are designed to detect, sensors can be located inside or outside of the facility or other asset. When an event is detected, the sensors use some type of communication method (such as wireless radio transmitters, conductors, or cables) to send signals to the control panel to generate the alarm. For example, a smoke detector sends a signal to a control panel when it detects smoke.

Alarms use either normally closed (NC) or normally open (NO) electric loops, or "circuits," to generate alarm signals. These two types of circuits are discussed separately below.

In NC loops or circuits, all of the system's sensors and switches are connected in series. The contacts are "at rest" in the closed (on) position, and current continually passes through the system. However, when an event triggers the sensor, the loop is opened, breaking the flow of current through the system and triggering the alarm. NC switches are used more often than are NO switches because the alarm will be activated if the loop or circuit is broken or cut, thereby reducing the potential for circumventing the alarm. This is known as a "supervised" system.

In NO loops or circuits, all of the system's sensors and switches are connected in parallel. The contacts are "at rest" in the open (off) position, and no current passes through the system. However, when an event triggers the sensor, the loop is closed. This allows current to flow through the loop, powering the alarm. NO systems are not supervised because the alarm will not be activated if the loop or circuit is broken or cut. However, adding an end-of-line resistor to a NO loop will cause the system to alarm if tampering is detected.

An arming station, which is the main user interface with the security system, allows the user to arm (turn on), disarm (turn off), and communicate with the system. How a specific system is armed will depend on how it is used. For example, while IDSs can be armed for continuous operation (24 hours a day), they are usually armed and disarmed according to the work schedule at a specific location so that personnel going about their daily activities do not set off the alarms. In contrast, fire protection systems are typically armed 24 hours a day.

A control panel receives information from the sensors and sends it to an appropriate location, such as a central operations station or a 24-hour monitoring facility. Once the alarm signal is received at the central monitoring location, personnel monitoring for alarms can respond (such as by sending security teams to investigate or by dispatching the fire department).

An annunciator responds to the detection of an event by emitting a signal. This signal may be visual, audible, electronic, or a combination of these three. For example, fire alarm signals will always be connected to audible annunciators, whereas intrusion alarms may not be.

Alarms can be reported locally, remotely, or both. A local alarm emits a signal at the location of the event (typically using a bell or siren). A local only alarm emits a signal at the location of the event but does not transmit the alarm signal to any other location (i.e., it does not transmit the alarm to a central monitoring location). Typically, the purpose of a local only alarm is to frighten away intruders, and possibly to attract the attention of someone who might notify the proper authorities. Because no signal is sent to a central monitoring location, personnel can only respond to a local alarm if they are in the area and can hear and/or see the alarm signal.

Fire alarm systems must have local alarms, including both audible and visual signals. Most fire alarm signal and response requirements are codified in the National Fire Alarm Code, National Fire Protection Association (NFPA) 72. NFPA 72 discusses the application, installation, performance, and maintenance of protective signaling systems and their components. In contrast to fire alarms, which require a local signal when fire is detected, many IDSs do not have a local alert device, because monitoring personnel do not wish to inform potential intruders that they have been detected. Instead, these types of systems silently alert monitoring personnel that an intrusion has been detected, thus allowing monitoring personnel to respond.

In contrast to systems that are set up to transmit local only alarms when the sensors are triggered, systems can also be set up to transmit signals to a central location, such as a control room or guard post at the utility or a police or fire station. Most fire/smoke alarms are set up to signal both at the location of the event and at a fire station or central monitoring station. Many insurance companies require that facilities install certified systems that include alarm communication to a central station. For example, systems certified by the Underwriters Laboratory (UL) require that the alarm be reported to a central monitoring station.

The main differences between alarm systems lie in the types of event detection devices used in different systems. Intrusion sensors, for example, consist of two main categories: perimeter sensors and interior (space) sensors. Perimeter intrusion sensors are typically applied on fences, doors, walls, windows, and so on, and are designed to detect an intruder before he or she accesses a protected asset (i.e., perimeter intrusion sensors are used to detect intruders attempting to enter through a door, window, etc.). In contrast, interior intrusion sensors are designed to detect an intruder who has already accessed the protected asset (i.e., interior intrusion sensors are used to detect intruders once they are already within a protected room or building). These two types of detection devices can be complementary, and they are often used together to enhance security for an asset. For example, a typical intrusion alarm system might employ a perimeter glass-break detector that protects against intruders accessing a room through a window, as well as an ultrasonic interior sensor that detects intruders that have gotten into the room without using the window. Table 9.6 lists and describes types of perimeter and interior sensors.

Fire detection/alarm systems consist of different types of fire detection devices and fire alarm systems available. These systems may detect fire, heat, smoke, or a combination of any of these. For example, a typical fire alarm system might consist of heat sensors, which are located throughout a facility and which detect high temperatures or a certain change in temperature over a fixed time period. A different system might be outfitted with both smoke and heat detection devices. A summary of several different types of fire/smoke/heat detection sensors is provided in Table 9.7.

Once a sensor in an alarm system detects an event, it must communicate an alarm signal. The two basic types of alarm communication systems are hardwired and wireless. Hardwired systems rely on wire that is run from the control panel to each of the detection devices and annunciators. Wireless systems transmit signals from a trans-

Table 9.6. Perimeter and Interior Sensors

Type of Perimeter Sensor	*Description*
Foil	Foil is a thin, fragile, lead-based metallic tape that is applied to glass windows and doors. The tape is applied to the window or door, and electric wiring connects this tape to a control panel. The tape functions as a conductor and completes the electric circuit with the control panel. When an intruder breaks the door or window, the fragile foil breaks, opening the circuit and triggering an alarm condition.
Magnetic switches (reed switches)	The most widely used perimeter sensor. They are typically used to protect doors, as well as windows that can be opened (windows that cannot be opened are more typically protected by foil alarms).
Glass break detectors	Placed on glass and sense vibrations in the glass when it is disturbed. The two most common types of glass break detectors are shock sensors and audio discriminators.
Type of Interior Sensor	*Description*
Passive infrared (PIR)	Presently the most popular and cost effective interior sensors. PIR detectors monitor infrared radiation (energy in the form of heat) and detect rapid changes in temperature within a protected area. Because infrared radiation is emitted by all living things, these types of sensors can be very effective.
Quad PIRs	Consist of two dual-element sensors combined in one housing. Each sensor has a separate lens and a separate processing circuitry, which allows each lens to be set up to generate a different protection pattern
Ultrasonic detectors	Emit high frequency sound waves and sense movement in a protected area by sensing changes in these waves. The sensor emits sound waves that stabilize and set a baseline condition in the area to be protected. Any subsequent movement within the protected area by a would-be intruder will cause a change in these waves, thus creating an alarm condition.
Microwave detectors	Emit ultrahigh frequency radio waves, and the detector senses any changes in these waves as they are reflected throughout the protected space. Microwaves can penetrate through walls, and thus a unit placed in one location may be able to protect multiple rooms.
Dual technology devices	Incorporate two different types of sensor technology (such as PIR and microwave technology) together in one housing. When both technologies sense an intrusion, an alarm is triggered.

Source: Water and Wastewater Security Product Guide, U.S. EPA, 2005.

Table 9.7. Fire/Smoke/Heat Detection Sensors

Detector Type	*Description*
Thermal detectors	Sense when temperatures exceed a set threshold (fixed temperature detectors) or when the rate of change of temperature increases over a fixed time period (rate-of-rise detectors).
Duct detectors	Located within the heating and ventilation ducts of the facility. These sensors detect the presence of smoke within the system's return or supply ducts. A sampling tube can be added to the detector to help span the width of the duct.
Smoke detectors	Sense invisible and/or visible products of combustion. The two principal types of smoke detectors are photoelectric and ionization detectors. The major differences between these devices are described below: • Photoelectric smoke detectors react to visible particles of smoke. These detectors are more sensitive to the cooler smoke with large smoke particles that is typical of smoldering fires. • Ionization smoke detectors are sensitive to the presence of ions produced by the chemical reactions that take place with few smoke particle, such as those typically produced by fast burning/flaming fires.
Multisensor detectors	A combination of photoelectric and thermal detectors. The photoelectric sensor serves to detect smoldering fires, while the thermal detector senses the heat given off from fast burning/flaming fires.
Carbon monoxide (CO) detectors	Used to indicate the outbreak of fire by sensing the level of carbon monoxide in the air. The detector has an electrochemical cell that senses carbon monoxide but not some other products of combustion.
Beam detectors	Designed to protect large, open spaces such as industrial warehouses. These detectors consist of three parts: the transmitter, which projects a beam of infrared light; the receiver, which registers the light and produces an electrical signal; and the interface, which processes the signal and generates alarm of fault signals. In the event of a fire, smoke particles obstruct the beam of light. Once a preset threshold is exceeded, the detector will go into alarm.
Flame detectors	Sense either ultraviolet (UV) or infrared (IR) radiation emitted by a fire.
Air-sampling detectors	Actively and continuously sample the air from a protected space and are able to sense the precombustion stages of incipient fire.

Source: Water and Wastewater Security Product Guide, U.S. EPA, 2005.

mitter to a receiver through the air—primarily using radio or other waves. Hardwired systems are usually lower cost, more reliable (they are not affected by terrain or environmental factors), and significantly easier to troubleshoot than wireless systems. However, a major disadvantage of hardwired systems is that it may not be possible to hardwire all locations (for example, it may be difficult to hardwire remote locations).

In addition, running wires to their required locations can be both time consuming and costly. The major advantage to using wireless systems is that they can often be installed in areas where hardwired systems are not feasible. However, wireless components can be much more expensive than hardwired systems. In addition, in the past, it has been difficult to perform self-diagnostics on wireless systems to confirm that they are communicating properly with the controller. Presently, the majority of wireless systems incorporate supervising circuitry, which allows the subscriber to know immediately if there is a problem with the system (such as a broken detection device or a low battery), or if a protected door or window has been left open.

Backflow Prevention Devices

Backflow prevention devices are designed to prevent backflow, which is the reversal of the normal and intended direction of water flow in a water system. Backflow is a potential problem in a water system because it can spread contaminated water back through a distribution system. For example, backflow at uncontrolled cross-connections (cross-connections are actual or potential connections between the public water supply and a source of contamination) or pollution can allow pollutants or contaminants to enter the potable water system. More specifically, backflow from private plumbing systems, industrial areas, hospitals, and other hazardous contaminant-containing systems into public water mains and wells poses serious public health risks and security problems. Cross-contamination from private plumbing systems can contain biological hazards (such as bacteria or viruses) or toxic substances that can contaminate and sicken an entire population. The majority of historical incidences of backflow have been accidental, but growing concern that contaminants could be intentionally backfed into a system is prompting increased awareness for private homes, businesses, industries, and areas most vulnerable to intentional strikes. Therefore, backflow prevention is a major tool for the protection of water systems.

Backflow may occur under two types of conditions: backpressure and backsiphonage. Backpressure is the reverse from normal flow direction within a piping system that is the result of the downstream pressure being higher than the supply pressure. These reductions in the supply pressure occur whenever the amount of water being used exceeds the amount of water supplied, such as during water line flushing, firefighting, or breaks in water mains. Backsiphonage is the reverse from normal flow direction within a piping system that is caused by negative pressure in the supply piping (i.e., the reversal of normal flow in a system caused by a vacuum or partial vacuum within the water supply piping). Backsiphonage can occur where there is a high velocity in a pipe line; when there is a line repair or break that is lower than a service point; or when there is lowered main pressure due to high water withdrawal rate, such as during firefighting or water main flushing.

Various types of backflow preventers are appropriate for use. The primary types of backflow preventers include the following:

- Air gap drains
- Double check valves
- Reduced pressure principle assemblies
- Pressure vacuum breakers

Biometric Security Systems

Biometrics involves measuring the unique physical characteristics or traits of the human body. Any aspect of the body that is measurably different from person to person—for example, fingerprints or eye characteristics—can serve as a unique biometric identifier for that individual. Biometric systems recognizing fingerprints, palm shape, eyes, face, voice, and signature compose the bulk of the current biometric systems, though systems that recognize other biological features do exist.

Biometric security systems use biometric technology combined with some type of locking mechanisms to control access to specific assets. In order to access an asset controlled by a biometric security system, an individual's biometric trait must be matched with an existing profile stored in a database. If there is a match between the two, the locking mechanism (which could be a physical lock, such as at a doorway, an electronic lock, such as at a computer terminal, or some other type of lock) is disengaged, and the individual is given access to the asset.

A biometric security system is typically composed of the following components:

- A sensor, which measures/records a biometric characteristic or trait
- A control panel, which serves as the connection point between various system components; communicates information back and forth between the sensor and the host computer; and controls access to the asset by engaging or disengaging the system lock based on internal logic and information from the host computer
- A host computer, which processes and stores the biometric trait in a database
- Specialized software, which compares an individual image taken by the sensor with a stored profile or profiles
- A locking mechanism, which is controlled by the biometric system
- A power source to power the system.

Biometric Hand and Finger Geometry Recognition

Hand and finger geometry recognition is the process of identifying an individual through the unique "geometry" (shape, thickness, length, width, etc.) of that individual's hand or fingers. Hand geometry recognition has been employed since the early 1980s and is among the most widely used biometric technologies for controlling access to important assets. It is easy to install and use and is appropriate for use in any location requiring use of two-finger highly accurate, nonintrusion biometric security. For example, it is currently used in numerous workplaces, day care facilities, hospitals, universities, airports, and power plants.

A newer option within hand geometry recognition technology is finger geometry recognition (not to be confused with fingerprint recognition). Finger geometry recognition relies on the same scanning methods and technologies as hand geometry recognition, but the scanner only scans two of the user's fingers, as opposed to his entire hand. Finger geometry recognition has been in commercial use since the mid 1990s and is mainly used in time and attendance applications (i.e., to track when individuals have entered and exited a location). To date, the only large-scale commercial use of two-finger geometry for controlling access is at Disney World, where season pass holders use the geometry of their index and middle finger to gain access to the facilities.

To use a hand or finger geometry unit, an individual presents his or her hand or fingers to the biometric unit for scanning. The scanner consists of a charged coupled device, which is essentially a high-resolution digital camera; a reflective platen on which the hand is placed; and a mirror or mirrors that help capture different angles of the hand or fingers. The camera scans individual geometric characteristics of the hand or fingers by taking multiple images while the user's hand rests on the reflective platen. The camera also captures depth, or three-dimensional information, through light reflected from the mirrors and the reflective platen. This live image is then compared to a template that was previously established for that individual when he was enrolled in the system. If the live scan of the individual matches the stored template, the individual is verified and given access to that asset. Typically, verification takes about two seconds. In access control applications, the scanner is usually connected to some sort of electronic lock, which unlocks the door, turnstile, or other entry barrier when the user is verified. The user can then proceed through the entrance. In time and attendance applications, the time that an individual checks in and out of a location is stored for later use.

As discussed above, hand and finger geometry recognition systems can be used in several different types of applications, including access control and time and attendance tracking. While time and attendance tracking can be used for security, it is primarily used for operations and payroll purposes (e.g., clocking in and clocking out). In contrast, access control applications are more likely to be security related. Biometric systems are widely used for access control and can be used on various types of assets, including entryways, computers, vehicles, and so on. However, because of their size, hand/finger recognition systems are primarily used in entryway access control applications.

Biometric Overview-Iris Recognition

The iris, which is the colored or pigmented area of the eye surrounded by the sclera (the white portion of the eye), is a muscular membrane that controls the amount of light entering the eye by contracting or expanding the pupil (the dark center of the eye). The dense, unique patterns of connective tissue in the human iris were first noted in 1936, but it was not until 1994, when algorithms for iris recognition were created and patented, that commercial applications using biometric iris recognition began to be used extensively. There are now two vendors producing iris recognition technology: both the original developer of these algorithms, as well as a second company, which has developed and patented a different set of algorithms for iris recognition.

The iris is an ideal characteristic for identifying individuals because it is formed in utero, and its unique patterns stabilize around eight months after birth. No two irises are alike; neither an individual's right or left irises, nor the irises of identical twins. The iris is protected by the cornea (the clear covering over the eye), and therefore it is not subject to the aging or physical changes (and potential variation) that are common to some other biometric measures, such as the hand, fingerprints, and the face. Although some limited changes can occur naturally over time, these changes generally occur in the iris' melanin and therefore affect only the eye's color, not its unique patterns. (In addition, because iris scanning uses only black and white images, color changes would not affect the scan anyway.). Thus, barring specific injuries or certain rare surgeries directly affecting the iris, the iris' unique patterns remain relatively unchanged over an individual's lifetime.

Iris recognition systems employ a monochromatic or black and white video camera that uses both visible and near infrared light to take video of an individual's iris. Video is used rather than still photography as an extra security procedure. The video is used to confirm the normal continuous fluctuations of the pupil as the eye focuses, which ensures that the scan is of a living human being, not a photograph or some other attempted hoax. A high-resolution image of the iris is then captured or extracted from the video, using a device often referred to as a "frame grabber." The unique characteristics identified in this image are then converted into a numeric code, which is stored as a template for that user.

Card Identification/Access/Tracking Systems

A card reader system is a type of electronic identification system that is used to identify a card and then perform an action associated with that card. Depending on the system, the card may identify where a person is or where they were at a certain time, or it may authorize another action, such as disengaging a lock. For example, a security guard may use his card at card readers located throughout a facility to indicate that he has checked a certain location at a certain time. The reader will store the information and/or send it to a central location, where it can be checked later to ensure that the guard has patrolled the area. Other card reader systems can be associated with a lock, so that the cardholder must have their card read and accepted by the reader before the lock disengages.

A complete card reader system typically consists of the following components:

- Access cards that are carried by the users
- Card readers, which read the card signals and send the information to control units
- Control units, which control the response of the card reader to the card
- A power source

A "card" may be a typical card or another type of device, such as a key fob or wand. These cards store electronic information, which can range from a simple code (e.g., the alphanumeric code on a Proximity card) to individualized personal data (e.g., biometric data on a Smartcard). The card reader reads the information stored on the card and sends it to the control unit, which determines the appropriate action to take when a card is presented. For example, in a card access system, the control unit compares the information on the card to stored access authorization information to determine if the card holder is authorized to proceed through the door. If the information stored in the card reader system indicates that the key is authorized to allow entrance through the doorway, the system disengages the lock and the key holder can proceed through the door.

There are many different types of card reader systems on the market. The primary differences between card reader systems are in the way that data is encoded on the cards, in the way these data are transferred between the card and the card reader, and in the types of applications for which they are best suited. However, all card systems are similar in the way that the card reader and control unit interact to respond to the card.

The different types of card reader systems include the following:

- Proximity
- Wiegand
- Smartcard
- Magnetic strip
- Bar code
- Infrared
- Barium ferrite
- Hollerith
- Mixed technologies

Table 9.8 summarizes various aspects of card reader technologies. The level of security (low, moderate, or high) is based on the level of technology a given card reader system has and how simple it is to duplicate that technology and thus bypass the security. Vulnerability ratings are based on whether the card reader can be damaged easily due to frequent use or difficult working conditions (e.g., weather conditions if the reader is located outside). Often this is influenced by the number of moving parts in the system—the more moving parts, the greater the system's potential susceptibility to damage. The life cycle rating is based on the durability of a given card reader system over its entire operational period. Systems requiring frequent physical contact between the reader and the card often have a shorter life cycle due to the wear and tear to which the equipment is exposed. For many card reader systems, the vulnerability rating and life cycle ratings have a reciprocal relationship. For instance, if a given system has a high vulnerability rating it will almost always have a shorter life cycle.

Card reader technology can be implemented for facilities of any size and with any number of users. However, because individual systems vary in the complexity of their technology and in the level of security they can provide to a facility, individual users must determine the appropriate system for their needs. Some important features to consider when selecting a card reader system include the following:

- The technological sophistication and security level of the card system
- The size and security needs of the facility
- The frequency with which the card system will be used (For systems that will experience a high frequency of use, it is important to consider a system that has a longer life cycle and lower vulnerability rating, thus making it more cost effective to implement.)
- The conditions in which the system will be used (e.g., will it be used on the interior or exterior of buildings, does it require light or humidity controls, etc. Most card reader systems can operate under normal environmental conditions, and therefore this would be a mitigating factor only in extreme conditions.)
- System costs

Exterior Intrusion-Buried Sensors

Buried sensors are electronic devices that are designed to detect potential intruders. The sensors are buried along the perimeters of sensitive assets and are able to detect intruder activity both aboveground and belowground. Some of these systems are composed of individual, stand-alone sensor units, while other sensors consist of buried cables.

There are four types of buried sensors that rely on different types of triggers: pressure or seismic, magnetic field, ported coaxial cable, and fiber-optic cables. These four

Table 9.8. Card Reader Technology

Types of Card Readers	*Technology*	*Life Cycle*	*Vulnerability*	*Level of Security*
Proximity	Embedded radio frequency circuits encoded with unique information	Long	Virtually none	Moderate to high
Wiegand	Short lengths of small-diameter, special alloy wire with unique magnetic properties	Long	Low susceptibility to damage; high durability due to embedded wires	Moderate to high
Magnetic stripe	Electromagnetic charges to encode information on a piece of tape attached to back of card	Moderate	Moderately susceptible to damage due to frequent use	Low to moderate
Bar code	Series of narrow and wide bars and spaces	Short	High; easily damaged	Low
Hollerith	Holes punched in a plastic or paper card and read optically	Short	High; easily damaged from frequent use	Low
Infrared	An encoded shadow pattern within the card, read using an infrared scanner	Moderate	IR scanners are optical and thus vulnerable to contamination	High
Barium ferrite	Uses small bits of magnetized barium ferrite, placed inside a plastic card; the polarity and location of the "spots" determines the coding	Moderate	Low susceptibility to damage; durable because spots are embedded in the material	Moderate to high
Smart cards	Patterns or series of narrow and wide bars and spaces	Short	High susceptibility to damage; low durability	Highest

Source: Water and Wastewater Security Product Guide, U.S. EPA, 2005.

sensors are all covert and terrain-following, meaning they are hidden from view and follow the contour of the terrain. The four types of sensors are described in more detail below. Table 9.9 presents the distinctions between the four types of buried sensors.

Exterior Intrusion Sensors

An exterior intrusion sensor is a detection device that is used in an outdoor environment to detect intrusions into a protected area. These devices are designed to detect an intruder, and then communicate an alarm signal to an alarm system. The alarm system can respond to the intrusion in many different ways, such as by triggering an audible or visual alarm signal or by sending an electronic signal to a central monitoring location that notifies security personnel of the intrusion.

Intrusion sensors can be used to protect many kinds of assets. Intrusion sensors that protect physical space are classified according to whether they protect indoor/interior space (e.g., an entire building or room within a building), or outdoor/exterior space (e.g., a fence line or perimeter). Interior intrusion sensors are designed to protect the interior space of a facility by detecting an intruder who is attempting to enter or who has already entered a room or building. In contrast, exterior intrusion sensors are designed to detect an intrusion into a protected outdoor/exterior area. Exterior protected areas are typically arranged as zones or exclusion areas placed so that the intruder is detected early in the intrusion attempt before the intruder can gain access to more valuable assets (e.g., into a building located within the protected area). Early detection creates additional time for security forces to respond to the alarm.

Exterior intrusion sensors are classified according to how the sensor detects the intrusion within the protected area. The three classes of exterior sensor technology include buried line sensors, fence-associated sensors, and freestanding sensors.

Buried Line Sensors

As the name suggests, buried line sensors are sensors that are buried underground and are designed to detect disturbances within the ground—such as disturbances

Table 9.9. Types of Buried Sensors

Type	*Description*
Pressure or seismic	Responds to disturbances in the soil
Magnetic field	Responds to a change in the local magnetic field caused by the movement of nearby metallic material
Ported coaxial cables	Responds to motion of a material with a high dielectric constant or high conductivity near the cables
Fiber-optic cables	Responds to a change in the shape of the fiber that can be sensed using sophisticated sensors and computer signal processing

Source: Adapted from Garcia, M. L., 2001.

caused by an intruder digging, crawling, walking, or running on the monitored ground. Because they sense ground disturbances, these types of sensors are able to detect intruder activity both on the surface and below ground. Individual types of exterior buried line sensors function in different ways, including by detecting motion, pressure, or vibrations within the protected ground or by detecting changes in some type of field (e.g., magnetic field) that the sensors generate within the protected ground. Specific types of buried line sensors include pressure or seismic sensors, magnetic field sensors, ported coaxial cables, and fiber-optic cables. Details on each of these sensor types are provided below.

- Buried line pressure or seismic sensors detect physical disturbances to the ground—such as vibrations or soil compression—caused by intruders walking, driving, digging, or otherwise physically contacting the protected ground. These sensors detect disturbances from all directions, and therefore can protect an area radially outward from their location; however, because detection may weaken as a function of distance from the disturbance, choosing the correct burial depth from the design area will be crucial. In general, sensors buried at a shallow depth protect a relatively small area but have a high probability of detecting intrusion within that area, while sensors buried at a deeper depth protect a wider area but have a lower probability of detecting intrusion into that area.
- Buried line magnetic field sensors detect changes in a local magnetic field that are caused by the movement of metallic objects within that field. This type of sensor can detect ferric metal objects worn or carried by an intruder entering a protected area on foot as well as vehicles being driven into the protected area.
- Buried line ported coaxial cable sensors detect the motion of any object (e.g., human body, metal, etc.) possessing high conductivity and located within close proximity to the cables. An intruder entering into the protected space creates an active disturbance in the electric field, thereby triggering an alarm condition.
- Buried line fiber-optic cable sensors detect changes in the attenuation of light signals transmitted within the cable. When the soil around the cable is compressed, the cable is distorted, and the light signal transmitted through the cable changes, initiating an alarm. This type of sensor is easy to install because it can be buried at a shallow burial depth (only a few centimeters) and still be effective.

Fence-Associated Sensors

Fence-associated sensors are either attached to an existing fence or installed in such a way as to create a fence. These sensors detect disturbances to the fence—such as those caused by an intruder attempting to climb the fence or by an intruder attempting to cut or lift the fence fabric. Exterior fence-associated sensors include fence disturbance sensors, taut-wire sensor fences, and electric field or capacitance sensors. Details on each of these sensor types are provided below.

- Fence disturbance sensors detect the motion or vibration of a fence, such as that caused by an intruder attempting to climb or cut through the fence. In general, fence disturbance sensors are used on chain-link fences or on other fence types where a moveable fence fabric is hung between fence posts.
- Taut-wire sensor fences are similar to fence disturbance sensors except that instead of attaching the sensors to a loose fence fabric, the sensors are attached to a wire that is stretched tightly across the fence. These types of systems are designed to detect changes in the tension of the wire rather than vibrations in the fence fabric. Taut-wire sensor fences can be installed over existing fences or as stand-alone fence systems.
- Electric field or capacitance sensors detect changes in capacitive coupling between wires that are attached to, but electrically isolated from, the fence. As opposed to other fence-associated intrusion sensors, both electric field and capacitance sensors generate an electric field that radiates out from the fence line, resulting in an expanded zone of protection relative to other fence-associated sensors, and allowing the sensor to detect intruders' presence before they arrive at the fence line. Proper spacing is necessary during installation of the electric field sensors to prevent a would-be intruder from slipping between largely spaced wires.

Free-standing Sensors

These sensors, which include active infrared, passive infrared, bistatic microwave, monostatic microwave, dual-technology, and video motion detection (VMD) sensors, consist of individual sensor units or components that can be set up in a variety of configurations to meet a user's needs. They are installed aboveground, and depending on how they are oriented relative to each other, they can be used to establish a protected perimeter or a protected space. More details on each of these sensor types are provided below.

- Active infrared sensors transmit infrared energy into the protected space and monitor for changes in this energy caused by intruders entering that space. In a typical application, an infrared light beam is transmitted from a transmitter unit to a receiver unit. If an intruder crosses the beam, the beam is blocked, and the receiver unit detects a change in the amount of light received, triggering an alarm. Different sensors can see single- and multiple-beam arrays. Single-beam infrared sensors transmit a single infrared beam. In contrast, multiple-beam infrared sensors transmit two or more beams parallel to each other. This multiple-beam sensor arrangement creates an infrared "fence."

- Passive infrared (PIR) sensors monitor the ambient infrared energy in a protected area and evaluate changes in that ambient energy that may be caused by intruders moving through the protected area. Detection ranges can exceed 100 yards on cold days with size and distance limitations dependent upon the background temperature. PIR sensors generate a nonuniform detection pattern (or "curtain") that has areas (or zones) of more sensitivity and areas of less sensitivity. The specific shape of the protected area is determined by the detector's lenses. The general shape common to many detection patterns is a series of long "fingers" emanating from the PIR and spreading in various directions. When intruders enter the detection area, the PIR sensor detects differences in temperature due to the intruder's body heat and triggers an alarm. While the PIR leaves unprotected areas between its fingers, an intruder would be detected if he passed from a nonprotected area to a protected area.
- Microwave sensors detect changes in received energy generated by the motion of an intruder entering a protected area. Monostatic microwave sensors incorporate a transmitter and a receiver in one unit, while bistatic sensors separate the transmitter and the receiver into different units. Monostatic sensors are limited to a coverage area of 400 feet, while bistatic sensors can cover an area up to 1,500 feet. For bistatic sensors, a zone of no detection exists in the first few feet in front of the antennas. This distance from the antennas to the point at which the intruder is first detected is known as the offset distance. Due to this offset distance, antennas must be configured so that they overlap one another (as opposed to being adjacent to each other), thereby creating long perimeters with a continuous line of detection.
- Dual-technology sensors consist of two different sensor technologies incorporated into one sensor unit. For example, a dual-technology sensor could consist of a passive infrared detector and a monostatic microwave sensor integrated into the same sensor unit.
- Video motion detection (VMD) sensors monitor video images from a protected area for changes in the images. Video cameras are used to detect unauthorized intrusion into the protected area by comparing the most recent image against a previously established one. Cameras can be installed on towers or other tall structures so that they can monitor a large area.

Fences

A fence is a physical barrier that can be set up around the perimeter of an asset. Fences often consist of individual pieces (such as individual pickets in a wooden fence or individual sections of a wrought iron fence) that are fastened together. Individual sections of the fence are fastened together using posts, which are sunk into the ground to provide stability and strength for the sections of the fence hung between them. Gates are installed between individual sections of the fence to allow access inside the fenced area.

Many fences are used as decorative architectural features to separate physical spaces from each other. They may also be used to physically mark the location of a boundary (such as a fence installed along a properly line). However, a fence can also serve as an effective means for physically delaying intruders from gaining access to a water or wastewater asset. For example, many utilities install fences around their primary facilities, around remote pump stations, or around hazardous materials storage areas or sensitive areas within a facility. Access to the area can be controlled through security at gates or doors through the fence (for example, by posting a guard at the gate or by locking it). In order to gain access to the asset, unauthorized persons would either have to go around or through the fence.

Fences are often compared with walls when determining the appropriate system for perimeter security. While both fences and walls can provide adequate perimeter security, fences are often easier and less expensive to install than walls. However, they do not usually provide the same physical strength that walls do. In addition, many types of fences have gaps between the individual pieces that make up the fence (i.e., the spaces between chain links in a chain link fence or the space between pickets in a picket fence). Thus, many types of fences allow the interior of the fenced area to be seen. This may allow intruders to gather important information about the locations or defenses of vulnerable areas within the facility.

There are numerous types of materials used to construct fences, including chain link iron, aluminum, wood, or wire. Some types of fences, such as split rails or pickets, may not be appropriate for security purposes because they are traditionally low and not physically strong. Potential intruders may be able to easily defeat these fences by jumping or climbing over them or by breaking through them. For example, the rails in a split fence may be broken easily.

Important security attributes of a fence include the height to which it can be constructed, the strength of the material comprising the fence, the method and strength of attaching the individual sections of the fence together at the posts, and the fence's ability to restrict the view of the assets inside the fence. Additional considerations should include the ease of installing the fence and the ease of removing and reusing sections

Table 9.10. Comparison of Different Fence Types

Specifications	*Chain Link*	*Iron*	*Wire (Wirewall)*	*Wood*
Height limitations	12′	12′	12′	8′
Installation requirements	Low	High	High	Low
Ability to remove/reuse	Low	High	Low	High
Ability to replace/repair	Medium	High	Low	High

Source: Water and Wastewater Security Product Guide, U.S. EPA, 2005.

of the fence. Table 9.10 provides a comparison of the important security and usability features of various fence types.

Some fences can include additional measures to delay or even detect potential intruders. Such measures may include the addition of barbed wire, razor wire, or other deterrents at the top of the fence. Barbed wire is sometimes employed at the base of fences as well. This can impede a would-be intruder's progress in even reaching the fence. Fences may also be fitted with security cameras to provide visual surveillance of the perimeter. Finally, some facilities have installed motion sensors along their fences to detect movement on the fence. Several manufacturers have combined these multiple perimeter security features into one product and offer alarms and other security features.

The correct implementation of a fence can make it a much more effective security measure. Security experts recommend the following when a facility constructs a fence:

- The fence should be at least seven to nine feet high.
- Any outriggers, such as bared wire, that are affixed on top of the fence should be angled out and away from the facility, and not in toward the facility. This will make climbing the fence more difficult and will prevent ladders from being placed against the fence.
- Other types of hardware can increase the security of the fence. This can include installing concertina wire along the fence (this can be done in front of the fence or at the top of the fence) or adding intrusion sensors, camera, or other hardware to the fence.
- All undergrowth should be cleared for several feet (typically six feet) on both sides of the fence. This will allow for a clearer view of the fence by any patrols in the area.
- Any trees with limbs or branches hanging over the fence should be trimmed so that intruders cannot use them to go over the fence. Also, it should be noted that fallen trees can damage fences, so management of trees around the fence can be important. This can be especially important in areas where a fence goes through a remote area.
- Fences that do not block the view from outside the fence to inside the fence allow patrols to see inside the fence without having to enter the facility.
- "No Trespassing" signs posted along a fence can be a valuable tool in prosecuting any intruders who claim that the fence was broken and that they did not enter through the fence illegally. Adding signs that highlight the local ordinances against trespassing can further deter simple troublemakers from illegally jumping/climbing the fence.

Films for Glass Shatter Protection

Most water and wastewater utilities have numerous windows on the outside of buildings, in doors, and in interior offices. In addition, many facilities have glass doors or other glass structures, such as glass walls or display cases. These glass objects are potentially vulnerable to shattering when heavy objects are thrown or launched at them, when explosions occur near them, or when there are high winds (for exterior glass). If the glass is shattered, intruders may potentially enter an area. In addition, shattered glass projected into a room from an explosion or from an object being thrown through a door or window can injure and potentially incapacitate personnel in the room. Materials that prevent glass from shattering can help to maintain the integrity of the door, window, or other glass object and can delay an intruder from gaining access. These materials can also prevent flying glass and thus reduce potential injuries.

Materials designed to prevent glass from shattering include specialized films and coatings. These materials can be applied to existing glass objects to improve their strength and ability to resist shattering. The films have been tested against many scenarios that could result in glass breakage, including penetration by blunt objects, bullets, high winds, and simulated explosions. Many vendors provide information on the results of these types of tests, and thus potential users can compare different product lines to determine which products best suit their needs.

The primary attributes of films for shatter protection include the following:

- The materials from which the film is made
- The adhesive that bonds the film to the glass surface
- The thickness of the film

Standard glass safety films are designed from high-strength polyester. Polyester provides both strength and elasticity, which is important for absorbing the impact of an object, spreading the force of the impact over the entire film, and resisting tearing. The polyester is also designed to be resistant to scratching, which can result when films are cleaned with abrasives or other industrial cleaners.

The bonding adhesive is important in ensuring that the film does not tear away from the glass surface. This can be especially important when the glass is broken so that the film does not peel off the glass and allow it to shatter. In addition, films applied to exterior windows can be subject to high concentrations of UV light, which can break down bonding materials.

Film thickness is measured in gauge or millimeters. According to test results reported by several manufacturers, film thickness appears to affect resistance to penetration/tearing, with thicker films being more resistant to penetration and tearing. However, the appreciation of a thicker film did not decrease glass fragmentation.

Many manufacturers offer films in different thicknesses. The standard film is usually one four-millimeter layer; thicker films are typically composed of several layers of the standard four-millimeter sheet. However, newer technologies have allowed the polyester to be microlayered to produce a stronger film without significantly increasing its thickness. In this microlayering process, each laminate film is composed of multiple microthin layers of polyester woven together at alternating angles. This provides increased strength for the film while maintaining the flexibility and thin profile of one film layer.

As described above, many vendors test their products in various scenarios that would lead to glass shattering, including simulated bomb blasts and simulation of the glass being struck by wind-blown debris. Some manufacturers refer to the Government Services Administration standard for bomb blasts, which requires resistance to tearing for a 4-psi blast. Other manufacturers use other measures and test for resistance to tearing. Many of these tests are not standard in that no standard testing or reporting methods have been adopted by any of the accepted standards-setting institutions. However, many of the vendors publish the procedure and the results of these tests on their websites, and this may allow users to evaluate the protectiveness of theses films. For example, several vendors evaluate the protectiveness of their films and the hazard resulting from blasts near windows with and without protective films. Protectiveness is usually evaluated based on the percentage of glass ejected from the window, and the height at which that ejected glass travels during the blast (for example, if the blasted glass tends to project upward into a room—potentially toward people's faces—it is a higher hazard than if it is blown downward into the room toward people's feet). There are some standard measures of glass breakage. For example, several vendors indicated that their products exceed the American Society for Testing and Materials Standard 64Z-95 - Standard Test Method for Glazing and Glazing Systems Subject to Air Blast Loadings. Vendors often compare unprotected glass to glass onto which their films have been applied in penetration or force tests, ballistic tests, or simulated explosions. Results generally show that applying films to the glass surfaces reduces breakage/penetration of the glass and can reduce the amount and direction of glass ejected from the frame. This in turn reduces the hazard from flying glass.

In addition to these types of tests, many vendors conduct standard physical tests on their products, such as tests for tensile and peel strength. Tensile strength indicates the strength per area of material, while the peel strength indicates the force it would take to peel the product from the glass surface. Several vendors indicate that their products exceed American National Standards Institute Standard Z97.1 for tensile strength and adhesion.

Vendors typically have a warranty against peeling or other forms of deterioration of their products. However, the warranty requires that the films be installed by manufacturer-certified technicians to ensure that they are applied correctly and therefore that the warranty is in effect. Warranties from different manufacturers may vary. Some may cover the cost of replacing the material only, while others include material plus installation. Because installation costs are significantly greater than material costs, different warranties may represent large differences in potential costs.

Fire Hydrant Locks

Fire hydrants are installed at strategic locations throughout a community's water distribution system to supply water for firefighting. However, because there are many hydrants in a system and they are often located in residential neighborhoods, industrial districts, and other areas where they cannot be easily observed and/or guarded, they are potentially vulnerable to unauthorized access. Many municipalities, states, and Environmental Protection Agency (EPA) regions have recognized this potential vulnerability and have instituted programs to lock hydrants. For example, EPA Region 1 has included locking hydrants as number seven on its "Drinking Water Security and Emergency Preparedness" Top Ten List for small groundwater suppliers.

A hydrant lock is a physical security device designed to prevent unauthorized access to the water supply through a hydrant. It can also ensure water and water pressure availability to firefighters and prevent water theft and associated lost water revenue. These locks have been successfully used in numerous municipalities and in various climates and weather conditions.

Fire hydrant locks are basically steel covers or caps that are locked in place over the operating nut of a fire hydrant. The lock prevents unauthorized persons from accessing the operating nut and opening the fire hydrant valve. The lock also makes it more difficult to remove the bolts from the hydrant and access the system that way. Finally, hydrant locks shield the valve from being broken off. Should a vandal attempt to breach the hydrant lock by force and succeed in breaking the hydrant lock, the vandal will only succeed in bending the operating valve. If the hydrant's operating valve is bent, the hydrant will not be operational, but the water asset remains protected and inaccessible to vandals. However, the entire hydrant will need to be replaced.

Hydrant locks are designed so that the hydrants can be operated by special key wrenches without removing the lock. These specialized wrenches are generally distributed to the fire department, public works department, and other authorized persons so that they can access the hydrants as needed. An inventory of wrenches and their serial numbers is generally kept by a municipality so that the location of all wrenches is known. These operating key wrenches may only be purchased by registered lock owners.

The most important features of hydrant locks are their strength and the security of their locking systems. Hydrant locks are constructed from stainless or alloyed steel. Stainless steel locks are stronger and are ideal for all climates; however, they are more expensive than alloy locks. The locking mechanisms for each fire hydrant locking system ensure that the hydrant can only be operated by authorized personnel who have the specialized key to work the hydrant.

Hatch Security

A hatch is basically a door that is installed on a horizontal plane (such as in a floor, a paved lot, or a ceiling), instead of on a vertical plane (such as in a building wall). Hatches are usually used to provide access to assets that are either located underground (such as hatches to basements or underground storage areas), or to assets located above ceilings (such as emergency roof exits). At water and wastewater facilities, hatches are typically used to provide access to underground vaults containing pumps, valves, or piping, or to the interior of water tanks or covered reservoirs. Securing a hatch by locking it or upgrading materials to give the hatch added strength can help to delay unauthorized access to any asset behind the hatch.

Like all doors, a hatch consists of a frame anchored to the horizontal structure, a door or doors, hinges connecting the door/doors to the frame, and a latching or locking mechanism that keeps the hatch door/doors closed.

It should be noted that improving hatch security is straightforward and that hatches with upgraded security features can be installed new or they can be retrofit for existing applications. Many municipalities already have specifications for hatch security at their water and wastewater utility assets.

Depending on the application, the primary security-related attributes of a hatch are the strength of the door and frame, its resistance to the elements and corrosion, its ability to be sealed against water or gas, and its locking features.

Hatches must be both strong and lightweight so that they can withstand typical static loads (such as people or vehicles walking or driving over them) while still being easy to open.

In addition, because hatches are typically installed at outdoor locations, they are usually designed from corrosion-resistant metal that can withstand the elements. Therefore, hatches are typically constructed from high gauge steel or lightweight aluminum.

Aluminum is typically the material of choice for hatches because it is lightweight and more corrosion resistant than steel. Aluminum is not as rigid as steel, so aluminum hatch doors may be reinforced with aluminum stiffeners to provide extra strength and rigidity. The doors are usually constructed from single or double layers (or "leaves") of material. Single-leaf designs are standard for smaller hatches, while double-leaf designs are required for larger hatches. In addition, aluminum products do not require painting. This is reflected in the warranties available with different products. Product warranties range from ten years to lifetime.

Steel is heavier per square foot than aluminum, and thus steel hatches will be heavier and more difficult to open than aluminum hatches of the same size. However, heavy steel hatch doors may have spring-loaded, hydraulic, or gas openers or other specialized features that help in opening the hatch and in keeping it open.

Many hatches are installed in outdoor areas, often in roadways or pedestrian areas. Therefore, the hatch installed for any given application must be designed to withstand the expected load at that location. Hatches are typically solid to withstand either pedestrian or vehicle loading. Pedestrian loading hatches are typically designed to withstand either 150 or 300 pounds per square feet of loading. The vehicle loading standard is the American Association of State Highway and Transportation Officials H-20 wheel loading standard of 16,000 pounds over an 8-inch by 20-inch area. It should be noted that these design parameters are for static loads and not dynamic loads; thus, the loading capabilities may not reflect potential resistance to other types of loads that may be more typical of an intentional threat, such as repeated blows from a sledge hammer or pressure generated by bomb blasts or bullets.

The typical design for a watertight hatch includes a channel frame that directs water away from the hatch. This can be especially important in a hatch on a storage tank because this will prevent liquid contaminants from being dumped on the hatch and leaking through into the interior. Hatches can also be constructed with gasket seals that are air, odor, and gas tight.

Typically, hatches for pedestrian-loading applications have hinges located on the exterior of the hatch, while hatches designed for H-20 loads have hinges located in the interior of the hatch. Hinges located on the exterior of the hatch may be able to be removed, thereby allowing intruders to remove the hatch door and access the asset behind the hatch. Therefore, installing H-20 hatches even for applications which do not require H-20 loading levels may increase security.

In addition to the location of the hinges, stock hinges can be replaced with heavy duty or security hinges that are more resistant to tampering.

The hatch locking mechanism is perhaps the most important part of hatch security. There are a number of locks that can be implemented for hatches, including the following:

- Slam locks (internal locks that are located within the hatch frame)
- Recessed cylinder locks
- Bolt locks
- Padlocks

Ladder Access Control

Water and wastewater utilities have a number of assets that are raised above ground level, including raised water tanks, raised chemical tanks, raised piping systems, and roof access points into buildings. In addition, communications equipment, antennae, or other electronic devices may be located on the top of these raised assets. Typically, these assets are reached by ladders that are permanently anchored to the asset. For example, raised water tanks typically are accessed by ladders that are bolted to one of the legs of the tank. Controlling access to these raised assets by controlling access to the ladder can increase security at a water or wastewater utility.

A typical ladder access control system consists of some type of cover that is locked or secured over the ladder. The cover can be a casing that surrounds most of the ladder, or a door or shield that covers only part of the ladder. In either case, several rungs of the ladder (the number of rungs depends on the size of the cover) are made inaccessible by the cover, and these rungs can only be accessed by opening or removing the cover. The cover is locked so that only authorized personnel can open or remove it and use the ladder. Ladder access controls are usually installed at several feet above ground level, and they usually extend several feet up the ladder so that they cannot be circumvented by someone accessing the ladder above the control system.

The important features of ladder access control are the size and strength of the cover and its ability to lock or otherwise be secured from unauthorized access.

The covers are constructed from aluminum or some type of steel. This should provide adequate protection from being pierced or cut through. The metals are corrosion resistant so that they will not corrode or become fragile from extreme weather conditions in outdoor applications. The bolts used to install each of these systems are galvanized steel. In addition, the bolts for each cover are installed on the inside of the unit so they cannot be removed from the outside.

Locks

A lock is a type of physical security device that can be used to delay or prevent a door, a window, a manhole, a filing cabinet drawer, or some other physical feature from being opened, moved, or operated. Locks typically operate by connecting two pieces together—such as by connecting a door to a door jamb or a manhole to its casement. Every lock has two modes—engaged (or locked), and disengaged (or opened). When a lock is disengaged, the asset on which the lock is installed can be accessed by anyone, but when the lock is engaged, only authorized persons have access to the locked asset.

Locks are excellent security features because they have been designed to function in many ways and to work on many different types of assets. Locks can also provide different levels of security depending on how they are designed and implemented. The security provided by a lock is dependent on several factors, including its ability to withstand physical damage (e.g., can it be cut off, broken, or otherwise physically disabled) as well as its requirements for supervision or operation (e.g., combinations may need to be changed frequently so that they are not compromised and the locks remain secure). While there is no single definition of the security of a lock, locks are often described as minimum, medium, or maximum security. Minimum security locks are those that can be easily disengaged (or "picked") without the correct key or code, or those that can be disabled easily (such as small padlocks that can be cut with bolt cutters). Higher security locks are more complex and thus are more difficult to pick or are sturdier and more resistant to physical damage.

Many locks only need to be unlocked from one side. For example, most door locks need a key to be unlocked only from the outside. A person opens such devices, called single-cylinder locks, from the inside by pushing a button or by turning a knob or handle. Double-cylinder locks require a key to be locked or unlocked from both sides.

Manhole Intrusion Sensors

Manholes are located at strategic locations throughout most municipal water, wastewater and other underground utility systems. Manholes are designed to provide access to the underground utilities, and therefore they are potential entry points to a system. For example, manholes in water or wastewater systems may provide access to sewer lines or vaults containing on/off or pressure-reducing water valves. Because many utilities run under other infrastructure (e.g., roads, buildings), manholes also provide potential access points to critical infrastructure as well as water and wastewater assets. In addition, because the portion of the system to which manholes provide entry is primarily located underground, access to a system through a manhole increases the chance that an intruder will not be seen. Therefore protecting manholes can be a critical component of guarding an entire community.

There are multiple methods for protecting manholes, including preventing unauthorized personnel from physically accessing the manhole and detecting attempts at unauthorized access to the manhole.

A manhole intrusion sensor is a physical security device designed to detect unauthorized access to the utility through a manhole. Monitoring a manhole that provides access to a water or wastewater system can mitigate two distinct types of threats. First, monitoring a manhole may detect access of unauthorized personnel to water or wastewater systems or assets through the manhole. Second, monitoring manholes may also allow the detection of the introduction of hazardous substances into the water system.

Several different technologies have been used to develop manhole intrusion sensors, including mechanical systems, magnetic systems, and fiber-optic and infrared sensors. Some of these intrusion sensors have been specifically designed for manholes, while others consist of standard, off-the-shelf intrusion sensors that have been implemented in a system specifically designed for application in a manhole.

Manhole Locks

A manhole lock is a physical security device designed to delay unauthorized access to the utility through a manhole. Locking a manhole that provides access to a water or wastewater system can mitigate two distinct types of threats. First, locking a manhole may delay access of unauthorized personnel to water or wastewater systems through the manhole. Second, locking manholes may also prevent the introduction of hazardous substances into the wastewater or storm water system.

Radiation Detection Equipment for Monitoring Personnel and Packages

One of the major potential threats facing water and wastewater facilities is contamination by radioactive substances. Radioactive substances brought onsite at a facility could be used to contaminate the facility, thereby preventing workers from safely entering the facility to perform necessary water treatment tasks. In addition, radioactive substances brought onsite at a water treatment plant could be discharged into the water source or the distribution system, contaminating the downstream water supply. Therefore, detection of radioactive substances being brought onsite can be an important security enhancement.

Different radionuclides have unique properties, and different equipment is required to detect different types of radiation. However, it is impractical and potentially unnecessary to monitor for specific radionuclides being brought onsite. Instead, for security purposes, it may be more useful to monitor for gross radiation as an indicator of unsafe substances.

In order to protect against these radioactive materials being brought onsite, a facility may be set up with monitoring sites outfitted with radiation detection instrumentation at entrances to the facility. Depending on the specific types of equipment chosen, this equipment would detect radiation emitted from people, packages, or other objects being brought through an entrance.

One of the primary differences between the different types of detection equipment is the means by which the equipment reads the radiation. Radiation may either be detected by direct measurement or through sampling.

Direct radiation measurement involves measuring radiation through an external probe on the detection instrumentation. Some direct measurement equipment detects radiation emitted into the air around the monitored object. Because this equipment detects radiation in the air, it does not require that the monitoring equipment make physical contact with the monitored object. Direct means for detecting radiation include using a walk-through portal-type monitor that would detect elevated radiation levels on a person or in a package, or by using a hand-held detector, which would be moved or swept over individual objects to locate a radioactive source.

Some types of radiation, such as alpha or low-energy beta radiation, have a short range and are easily shielded by various materials. These types of radiation cannot be measured through direct measurement. Instead, they must be measured through sampling. Sampling involves wiping the surface to be tested with a special filter cloth and then reading the cloth in a special counter. For example, specialized smear counters measure alpha and low-energy beta radiation.

Reservoir Covers

Reservoirs are used to store raw or untreated water. They can be located underground (buried), at ground level, or on an elevated surface. Reservoirs can vary significantly in size; small reservoirs can hold as little as 1,000 gallons, while larger reservoirs may hold many millions of gallons.

Reservoirs can be either natural or man-made. Natural reservoirs can include lakes or other contained water bodies, while man-made reservoirs usually consist of some sort of engineered structure, such as a tank or other impoundment structure. In addition to the water containment structure itself, reservoir systems may also include associated water treatment and distribution equipment, including intakes, pumps, pump houses, piping systems, chemical treatment and chemical storage areas, and so on.

Drinking water reservoirs are of particular concern because they are potentially vulnerable to contamination of the stored water, either through direct contamination of the storage area or through infiltration of the equipment, piping, or chemicals associated with the reservoir. For example, because many drinking water reservoirs are designed as aboveground, open-air structures, they are potentially vulnerable to airborne deposition, bird and animal wastes, human activities, and dissipation of chlorine or other treatment chemicals. However, one of the most serious potential threats to the system is direct contamination of the stored water through dumping contaminants into the reservoir. Utilities have taken various measures to mitigate this type of threat, including fencing off the reservoir, installing cameras to monitor for intruders, and monitoring for changes in water quality. Another option for enhancing security is covering the reservoir using some type of manufactured cover to prevent intruders from gaining physical access to the stored water. Implementing a reservoir cover may or may not be practical depending on the size of the reservoir (for example, covers are not typically used on natural reservoirs because they are too large for the cover to be technically feasible and cost effective). This section will focus on drinking water reservoir covers, where and how they are typically implemented, and how they can be used to reduce the threat of contamination of the stored water. While covers can enhance the reservoir's security, it should be noted that covering a reservoir typically changes the reservoir's operational requirements. For example, vents must be installed in the cover to ensure gas exchange between the stored water and the atmosphere.

A reservoir cover is a structure installed on or over the surface of the reservoir to minimize water quality degradation. The three basic design types for reservoir covers are floating, fixed, and air supported.

A variety of materials are used when manufacturing a cover, including reinforced concrete, steel, aluminum, polypropylene, chlorosulfonated polyethylene, or ethylene interpolymer alloys. The following are several factors that affect a reservoir cover's effectiveness:

- The location, size, and shape of the reservoir
- The ability to lay/support a foundation (for example, footing, soil, and geotechnical support conditions)
- The length of time the reservoir can be removed from service for cover installation or maintenance
- Aesthetic considerations
- Economic factors, such as capital and maintenance costs

For example, it may not be practical to install a fixed cover over a reservoir if the reservoir is too large or if the local soil conditions cannot support a foundation. A floating or air-supported cover may be more appropriate for these types of applications.

In addition to the practical considerations for installation of these types of covers, there are a number of operations and maintenance concerns that affect the utility of a cover for specific applications, including how different cover materials will withstand local climatic conditions, what types of cleaning and maintenance will be required for each particular type of cover, and how these factors will affect the cover's lifespan and its ability to be repaired when it is damaged.

The primary feature affecting the security of a reservoir cover is its ability to maintain its integrity. Any type of cover, no matter what its construction material, will provide good protection from contamination by rainwater or atmospheric deposition, as well as from intruders attempting to access the stored water with the intent of causing intentional contamination. The covers are large and heavy, and it is difficult to circumvent them to get into the reservoir. At the very least, it would take a determined intruder, as opposed to a vandal, to defeat the cover.

Passive Security Barriers

One of the most basic threats facing any facility is intruders accessing the facility with the intention of causing damage to its assets. These threats may include intruders actually entering the facility as well as intruders attacking the facility from outside without actually entering it (e.g., detonating a bomb near enough to the facility to cause damage within its boundaries).

Security barriers are one of the most effective ways to counter the threat of intruders accessing a facility or the facility perimeter. Security barriers are large, heavy structures that are used to control access through a perimeter by either vehicles or personnel. They can be used in many different ways depending on how/where they are located at the facility. For example, security barriers can be used on or along driveways or roads to direct traffic to a checkpoint (e.g., a facility may install jersey barriers in a road to direct traffic in a certain direction). Other types of security barriers (crash beams, gates) can be installed at the checkpoint so that guards can regulate which vehicles can access the facility. Finally, other security barriers (e.g., bollards or security planters) can be used along the facility perimeter to establish a protective buffer area between the facility and approaching vehicles. Establishing such a protective buffer can help in mitigating the effects of the type of bomb blast described above, both by potentially absorbing some of the blast, and also by increasing the "stand-off" distance between the blast and the facility. (The force of an explosion is reduced as the shock wave travels farther from the source, and thus the farther the explosion is from the target, the less effective it will be in damaging the target.)

Security for Doorways: Side-Hinged Doors

Doorways are the main access points to a facility or to rooms within a building. They are used on the exterior or in the interior of buildings to provide privacy and security for the areas behind them. Different types of doorway security systems may be installed in different doorways depending on the needs or requirements of the buildings or rooms. For example, exterior doorways tend to have heavier doors to withstand the elements and to provide some security to the entrance of the building. Interior doorways in office areas may have lighter doors that may be primarily designed to provide privacy rather than security. Therefore, these doors may be made of glass or lightweight wood. Doorways in industrial areas may have sturdier doors than other interior doorways and may be designed to provide protection or security for areas behind the doorway. For example, fireproof doors may be installed in chemical storage areas or in other areas where there is a danger of fire.

Because they are the main entries into a facility or a room, doorways are often prime targets for unauthorized entry into a facility or an asset. Therefore, securing doorways may be a major step in providing security at a facility.

A doorway includes the following four main components:

- The door, which blocks the entrance. The primary threat to the actual door is breaking or piercing through the door. Therefore, the primary security features of doors are their strength and resistance to various physical threats, such as fire or explosions.
- The door frame, which connects the door to the wall. The primary threat to a door frame is that the door can be pried away from the frame. Therefore, the primary security feature of a door frame is its resistance to prying.
- The hinges, which connect the door to the door frame. The primary threat to door hinges is that they can be removed or broken, which will allow intruders to remove the entire door. Therefore, security hinges are designed to be resistant to breaking. They may also be designed to minimize the threat of removal from the door.

- The lock, which connects the door to the door frame. Use of the lock is controlled through various security features, such as keys and combinations, so that only authorized personnel can open the lock and go through the door. Locks may also incorporate other security features, such as software or other systems to track overall use of the door or to track individuals using the door.

Each of these components is integral in providing security for a doorway, and upgrading the security of only one of these components while leaving the other components unprotected may not increase the overall security of the doorway. For example, many facilities upgrade door locks as a basic step in increasing the security of a facility. However, if the facilities do not also focus on increasing security for the door hinges or the door frame, the door may remain vulnerable to being removed from its frame, thereby defeating the increased security of the door lock.

The primary attribute for the security of a door is its strength. Many security doors are 4–20-gauge hollow metal doors consisting of steel plates over a hollow cavity reinforced with steel stiffeners to give the door extra stiffness and rigidity. This increases resistance to blunt force used to try to penetrate through the door. The space between the stiffeners may be filled with specialized materials to provide fire, blast, or bullet resistance to the door.

The Windows and Doors Manufacturers Association has developed a series of performance attributes for doors. These include the following:

- Structural resistance
- Forced entry resistance
- Hinge style screw resistance
- Split resistance
- Hinge resistance
- Security rating
- Fire resistance
- Bullet resistance
- Blast resistance

The first five bullets provide information on a door's resistance to standard physical breaking and prying attacks. These tests are used to evaluate the strength of the door and the resistance of the hinges and the frame in a standardized way. For example, the Rack Load Test simulates a prying attack on a corner of the door. A test panel is restrained at one end, and a third corner is supported. Loads are applied and measured at the fourth corner. The Door Impact Test simulates a battering attack on a door and frame using impacts of 200 foot pounds by a steel pendulum. The door must remain fully operable after the test. It should be noted that door glazing is also rated for resistance to shattering and other damage. Manufacturers will be able to provide security ratings for these features of a door as well.

Door frames are an integral part of doorway security because they anchor the door to the wall. Door frames are typically constructed from wood or steel, and they are installed such that they extend for several inches over the doorway that has been cut into the wall. For added security, frames can be designed to have varying degrees of overlap with, or wrapping over, the underlying wall. This can make prying the frame from the wall more difficult. A frame formed from a continuous piece of metal (as opposed to a frame constructed from individual metal pieces) will prevent prying between pieces of the frame.

Many security doors can be retrofit into existing frames; however, many security door installations include replacing the door frame as well as the door itself. For example, bullet-resistance per UL 752 requires resistance of the door and frame assembly, and thus replacing the door only would not meet UL 752 requirements.

Valve Lockout Devices

Valves are utilized as control elements in water and wastewater process piping networks. They regulate the flow of both liquids and gases by opening, closing, or obstructing a flow passageway. Valves are typically located where flow control is necessary. They can be located in-line or at pipeline and tank entrance and exit points. They can serve multiple purposes in a process pipe network, including the following:

- Redirecting and throttling flow
- Preventing backflow
- Shutting off flow to a pipeline or tank (for isolation purposes)
- Releasing pressure
- Draining extraneous liquid from pipelines or tanks
- Introducing chemicals into the process network
- Serving as access points for sampling process water

Valves are located at critical junctures throughout water and wastewater systems, both onsite at treatment facilities and offsite within water distribution and wastewater collection systems. They may be located either aboveground or belowground. Because many valves are located within the community, it is critical to provide protection against valve tampering. For example, tampering with a pressure relief valve could result in a pressure buildup and potential explosion in the piping network. On a larger scale, addition of a pathogen or chemical to the water distribution system through an unprotected valve could result in the release of that contaminant to the general population.

Security products available to protect aboveground valves are different from those for belowground valves. For example, valve lockout devices can be purchased to protect valves and valve controls located aboveground. Vaults containing underground valves can be locked to prevent access to these valves.

As described above, a lockout device can be used as a security measure to prevent unauthorized access to aboveground valves located within water and wastewater systems. Valve lockout devices are locks that are specially designed to fit over valves and valve handles to control their ability to be turned or seated. These devices can be used to lock the valve into the desired position. Once the valve is locked, it cannot be turned unless the locking device is unlocked or removed by an authorized individual.

Various valve lockout options are available for municipal and industrial use, including cable lockouts, padlocked chains/cables, and valve-specific lockouts.

Many of these lockout devices are not specifically designed for use in the water/wastewater industry (e.g., chains, padlocks), and are available from a local hardware store or manufacturer specializing in safety equipment. Other lockout devices (e.g., valve-specific lockouts or valve box locks) are more specialized and must be purchased from safety or valve-related equipment vendors.

The three most common types of valves for which lockout devices are available are gate, ball, and butterfly valves. Each is described in more detail below.

- Gate valve lockouts are designed to fit over the operating hand wheel of the gate valve to prevent it from being turned. The lockout is secured in place with a padlock. Two types of gate valve lockouts are available: diameter specific and adjustable. Diameter-specific lockouts are available for handles ranging from one to thirteen inches in diameter. Adjustable gate valve lockouts can be adjusted to fit any handle ranging from one to six or more inches in diameter.
- Ball valve lockouts are available in several different configurations, all of which are designed to prevent rotation of the valve handle. The three major configurations available are a wedge shape for one- to three-inch valves, a lockout that completely covers 3/8- to eight-inch ball valve handles, and a universal lockout that can be applied to quarter turn valves of varying sizes and geometric handle dimensions. All three types of ball valve lockouts can be installed by sliding the lockout device over the ball valve handle and securing it with a padlock.
- Butterfly valve lockouts function in a similar manner to the ball valve lockouts. The polypropylene lockout device is placed over the valve handle and secured with a padlock. This type of lockout has been commonly used in the bottling industry.

A major difference between valve-specific lockout devices and the padlocked chain or cable lockouts discussed earlier is that the former does not need to be secured to an anchoring device in the floor or the piping system. In addition, valve-specific lockouts eliminate potential tripping or access hazards that may be caused by chains or cable lockouts applied to valves located near walkways or frequently maintained equipment.

Valve-specific lockout devices are available in a variety of colors, which can be useful in distinguishing different valves. For example, different colored lockouts can be used to distinguish the type of liquid passing through the valve (e.g., treated, untreated, potable, chemical) or to identify the party responsible for maintaining the lockout. Implementing a system of different colored locks on operating valves can increase system security by reducing the likelihood of an operator inadvertently opening the wrong valve and causing a problem in the system.

Security for Vents

Vents are installed in aboveground covered water reservoirs and in underground reservoirs to allow ventilation of the stored water. Specifically, vents permit the passage of air that is being displaced from, or drawn into, the reservoir as the water level in the reservoir rises and falls due to system demands. Small reservoirs may require only one vent, whereas larger reservoirs may have multiple vents throughout the system.

While the specific vent design for any given application will vary depending on the design of the reservoir, every vent consists of an open-air connection between the reservoir and the outside environment. Although these air exchange vents are an integral part of covered or underground reservoirs, they also represent a potential security threat. Improving vent security by making the vents tamper resistant or by adding other security features, such as security screens or security covers, can enhance the security of the entire water system.

Many municipalities already have specifications for vent security at their water assets. These specifications typically include the following requirements:

- Vent openings are to be angled down or shielded to minimize the entrance of surface and/or rainwater into the vent through the opening.
- Vent designs are to include features to obstruct insects, birds, other animals, and dust.
- Corrosion-resistant materials are to be used to construct the vents.

Some states have adopted more specific requirements for added vent security at their water utility assets. For example, Utah's Department of Environmental Quality, Division of Drinking Water, Division of Administrative Rules, provides specific requirements for public drinking water storage tanks. The rules for drinking water storage tanks as they apply to venting are set forth in Utah-R309-545-15: Venting, and include the following requirements:

- Drinking water storage tank vents must have an open discharge on buried structures. The vents must be located 24 to 36 inches above the earthen covering.
- The vents must be located and sized to avoid blockage during winter conditions.

In a second example, Washington state's "Drinking Water Tech Tips: Sanitary Protection of Reservoirs" document states that vents must be protected to prevent the water supply from being contaminated. The document indicates that noncorrodible No. 4 mesh may be used to screen vents on elevated tanks. The document continues to state that the vent opening for storage facilities located underground or at ground level should be 24 to 36 inches above the roof or ground and that it must be protected with a No. 24 mesh non-corrodible screen. New Mexico's administrative code also specifies that vents must be covered with No. 24 mesh (NMAC Title 20, Chapter 7, Subpart I, 208.E).

As described above, Washington and New Mexico, as well as many other municipalities, require vents to be screened using a noncorrodible mesh to minimize the entry of insects, other animals, and rain-borne contamination into the vents. When selecting the appropriate mesh size, it is important to identify the smallest mesh size that meets both the strength and the durability requirements for that application.

Visual Surveillance Monitoring

Visual surveillance is used to detect threats through continuous observation of important or vulnerable areas of an asset. The observations can also be recorded for later review or use (for example, in court proceedings). Visual surveillance systems can be used to monitor various parts of collection, distribution, or treatment systems, including the perimeter of a facility, outlying pumping stations, or entry or access points into specific buildings. These systems are also useful in recording individuals who enter or leave a facility, thereby helping to identify unauthorized access. Images can be transmitted live to a monitoring station, where they can be monitored in real time, or they can be recorded and reviewed later. Many facilities have found that a combination of electronic surveillance and security guards provides an effective means of facility security.

Visual surveillance is provided through a CCTV system, in which the capture, transmission, and reception of an image is localized within a closed circuit. This is different from other broadcast images, such as over-the-air television, which is broadcast over the air to any receiver within range.

At a minimum, a CCTV system consists of one or more cameras, a monitor for viewing the images, and a system for transmitting the images from the camera to the monitor. Specific attributes and features of camera systems, lenses, and lighting systems are presented in Table 9.11.

Table 9.11. Attributes of Camera Systems, Lenses, and Lighting Systems

Camera Systems	
Attribute	*Discussion*
Camera type	Major factors in choosing the correct camera are the resolution of the image required and lighting of the area to be viewed. • Solid State (including charge coupled device, charge priming device, charge injection device, and metal oxide substrate): These cameras are becoming predominant in the marketplace because of their high resolution and their elimination of problems inherent in tub cameras. • Thermal: These cameras are designed for night vision. They require no light and use differences in temperature between objects in the field of view to produce a video image. Resolution is low compared to other cameras, and the technology is currently expensive relative to other technologies. • Tube: These cameras can provide high resolution but burn out and must be replaced after one to two years. In addition, tube performance can degrade over time. Finally, tube cameras are prone to burn images in the tube replacement.
Resolution (the ability to see fine details)	User must determine the amount of resolution depending on the level of detail required for threat determination. A high-definition focus with a wide field of vision will give an optimal viewing area.
Field of vision width	Cameras are designed to cover a defined field of vision, which is usually defined in degrees. The wider the field of vision, the more area a camera will be able to monitor.
Type of image produced (color, black and white, thermal)	Color images may allow the identification of distinctive markings, while black and white images may provide sharper contrast. Thermal imaging allows the identification of heat sources (such as human beings or other living creatures) from low-light environments; however, thermal images are not effective in identifying specific individuals (e.g., for subsequent legal processes).
Pan/tilt/zoom (PTZ)	Panning (moving the camera in a horizontal plane), tilting (moving the camera in a vertical plane), and zooming (moving the lens to focus on objects that are at different distances from the camera) allow the camera to follow a moving object. Different systems allow these functions to be controlled manually or automatically. Factors to be considered in PTZ cameras are the degree of coverage for pan and tilt functions and the power of the zoom lens.
Lenses	
Format	Lens format determines the maximum image size to be transmitted.
Focal length	This is the distance from the lens to the center of the focus. The greater the focal length, the higher the magnification, but the narrower the field of vision.
F number	F number is the ability to gather light. Smaller F numbers may be required for outdoor applications where light cannot be controlled as easily.
Distance and width approximation	The distance and width approximations are used to determine the geometry of the space that can be monitored at the best resolution.
Lighting Systems	
Intensity	Light intensity must be great enough for the camera type to produce sharp images. Light can be generated from natural or artificial sources. Artificial sources can be controlled to produce the amount and distribution of light required for a given camera and lens.
Evenness	Light must be distributed evenly over the field of view so that there are no darker or shadowy areas. If there is a high contrast between lighter and darker areas, brighter areas may appear washed out (i.e., details cannot be distinguished) while no specific objects can be viewed from darker areas.
Location	Light sources must be located above the camera so that light does not shine directly into the camera.

Source: Water and Wastewater Security Product Guide, U.S. EPA, 2005.

WATER MONITORING DEVICES

Earlier it was pointed out that proper security preparation really comes down to a three-legged approach: detect, delay, respond. The first leg of security, to detect, is discussed in this section. Specifically, this section deals with the monitoring of water samples to detect toxicity and/or contamination. The following major monitoring tools, which can be used to identify anomalies in process streams or finished water that may represent potential threats, are discussed:

- Sensors for monitoring chemical, biological, and radiological contamination
- Chemical sensor - Arsenic measurement system
- Chemical sensor for toxicity (adapted biochemical oxygen demand [BOD] analyzer)
- Chemical sensor - total organic carbon analyzer
- Chemical sensor - chlorine measurement system
- Chemical sensor - portable cyanide analyzer
- Portable field monitors to measure volatile organic compounds (VOCs)
- Radiation detection equipment
- Radiation detection equipment for monitoring water assets
- Toxicity monitoring/toxicity meters

Sensors for Monitoring Chemical, Biological, and Radiological Contamination

Toxicity tests measure water toxicity by monitoring adverse biological effects on test organisms. Toxicity tests have traditionally been used to monitor wastewater effluent streams for National Pollutant Discharge Elimination System (NPDES) permit compliance or to test water samples for toxicity. However, this technology can also be used to monitor drinking water distribution systems or other water/wastewater streams for toxicity. Currently, several types of biosensors and toxicity tests are being adapted for use in the water/wastewater security field. The keys to using biomonitoring or biosensors for drinking water or other water/wastewater asset security are rapid response and the ability to use the monitor at critical locations in the system, such as in water distribution systems downstream of pump stations or prior to the biological process in a wastewater treatment plant. While there are several different organisms that can be used to monitor for toxicity (including bacteria, invertebrates, and fish), bacteria-based biosensors are ideal for use as early warning screening tools for drinking water security because bacteria usually respond to toxics in a matter of minutes. In contrast to methods using bacteria, toxicity screening methods that use higher-level organisms such as fish may take several days to produce a measurable result. Bacteria-based biosensors have recently been incorporated into portable instruments, making rapid response and field testing practical. These portable meters detect decreases in biological activity (e.g., decreases in bacterial luminescence), which are highly correlated with increased levels of toxicity.

At the present time, few utilities are using biologically based toxicity monitors to monitor water/wastewater assets for toxicity, and very few products are now commercially available. Several new approaches to the rapid monitoring of microorganisms for security purposes (e.g., microbial source tracking) have been identified. However, most of these methods are still in the research and development phase.

Sensors for Monitoring Chemical, Biological, and Radiological Contamination

Water quality monitoring sensor equipment may be used to monitor key elements of water or wastewater treatment processes (such as influent water quality, treatment processes, or effluent water quality) to identify anomalies that may indicate threats to the system. Some sensors, such as sensors for biological organisms or radiological contaminants, measure potential contamination directly, while others, particularly some chemical monitoring systems, measure "surrogate" parameters that may indicate problems in the system but do not identify sources of contamination directly. In addition, sensors can provide more accurate control of critical components in water and wastewater systems and may provide a means of early warning so that the potential effects of certain types of attacks can be mitigated. One advantage of using chemical and biological sensors to monitor for potential threats to water and wastewater systems is that many utilities already employ sensors to monitor potable water (raw or finished) or influent/effluent for Safe Drinking Water Act (SDWA) or Clean Water Act water quality compliance or process control.

Chemical sensors that can be used to identify potential threats to water and wastewater systems include inorganic monitors (e.g., chlorine analyzer), organic monitors (e.g., total organic carbon analyzer), and toxicity meters. Radiological meters can be used to measure concentrations of several different radioactive species. Monitors that use biological species can be used as sentinels for the presence of contaminants of concern, such as toxics. At the present time, biological monitors are not in widespread use and very few biomonitors are used by drinking water utilities in the United States.

Monitoring can be conducted using either portable or fixed-location sensors. Fixed-location sensors are usually used as part of a continuous, on-line monitoring system. Continuous monitoring has the advantage of enabling immediate notification when there is an upset. However, the sampling points are fixed and only certain points in the system can be monitored. In addition, the number of monitoring locations needed to capture the physical, chemical, and biological complexity of a system can be prohibitive. The use of portable sensors can overcome this problem of monitoring many points in the system. Portable sensors can be used to analyze grab samples at any point in the system but have the disadvantage that they provide measurements only at one point in time.

Chemical Sensor—Arsenic Measurement System

Arsenic is an inorganic toxin that occurs naturally in soils. It can enter water supplies from many sources, including erosion of natural deposits, runoff from orchards, runoff from glass and electronics production wastes, or leaching from products treated with arsenic, such as wood. Synthetic organic arsenic is also used in fertilizer.

Arsenic toxicity is primarily associated with inorganic arsenic. Ingestion has been linked to cancerous health effects, including cancer of the bladder, lungs, skin, kidney, nasal passages, liver, and prostate. Arsenic ingestion has also been linked to noncancerous cardiovascular, pulmonary, immunological, neurological, and endocrine problems. According to the EPA's SDWA Arsenic Rule, inorganic arsenic can exert toxic effects after acute (short-term) or chronic (long-term) exposure. Toxicological data for acute exposure, which is typically given as an LD50 value (the dose that would be lethal to 50 percent of the test subjects in a given test), suggests that the LD50 of arsenic ranges from one to four milligrams of arsenic per kilogram of body weight. This dose would correspond to a lethal dose range of 70 to 280 milligrams for 50 percent of adults weighing 70 kilograms. At nonlethal but high acute doses, inorganic arsenic can cause gastroenterological effects, shock, neuritis (continuous pain), and vascular effects in humans. The EPA has set a maximum contaminant level goal of 0 for arsenic in drinking water; the current enforceable maximum contaminant level (MCL) is 0.050 mg/L. As of January 23, 2006, the enforceable MCL for arsenic will be 0.010 mg/L.

The SDWA requires arsenic monitoring for public water systems. The Arsenic Rule indicates that surface water systems must collect one sample annually; groundwater systems must collect one sample in each compliance period (once every three years). Samples are collected at entry points to the distribution system, and analysis is done in the lab using one of several EPA-approved methods, including Inductively Coupled Plasma Mass Spectroscopy (ICP-MS, U.S. EPA 200.8) and several atomic absorption methods. However, several different technologies, including colorimetric test kits and portable chemical sensors, are currently available for monitoring inorganic arsenic concentrations in the field. These technologies can provide a quick estimate of arsenic concentrations in a water sample. Thus, these technologies may be useful for spot-checking different parts of a drinking water system (e.g., reservoirs, isolated areas of distribution systems) to ensure that the water is not contaminated with arsenic.

Chemical Sensor—Total Organic Carbon Analyzer

One manufacturer has adapted a BOD analyzer to measure oxygen consumption as a surrogate for general toxicity. The critical element in the analyzer is the bioreactor, which is used to continuously measure the respiration of the biomass under stable conditions. As the toxicity of the sample increases, the oxygen consumption in the sample decreases. An alarm can be programmed to sound if oxygen reaches a minimum concentration (i.e., if the sample is strongly toxic). The operator must then interpret the results into a measure of toxicity.

Note that at the current time, it is difficult to directly define the sensitivity and/or the detection limit of toxicity measurement devices because limited data is available regarding specific correlation of decreased oxygen consumption and increased toxicity of the sample.

Chemical Sensor—Total Organic Carbon Analyzer

Total Organic Carbon (TOC) analysis is a well-defined and commonly used methodology that measures the carbon content of dissolved and particulate organic matter present in water. Many water utilities monitor TOC to determine raw water quality or to evaluate the effectiveness of processes designed to remove organic carbon. Some wastewater utilities also employ TOC analysis to monitor the efficiency of the treatment process. In addition to these uses for TOC monitoring, measuring changes in TOC concentrations can be an effective surrogate for detecting contamination from organic compounds (e.g., petrochemicals, solvents, pesticides). Thus, while TOC analysis does not give specific information about the nature of the threat, identifying changes in TOC can be a good indicator of potential threats to a system.

TOC analysis consists of inorganic carbon removal oxidation of the organic carbon into carbon dioxide and quantification of the carbon dioxide. The primary differences between different on-line TOC analyzers are in the methods used for oxidation and carbon dioxide quantification.

The oxidation step can be high or low temperature. The determination of the appropriate analytical method (and thus the appropriate analyzer) is based on the expected characteristics of the wastewater sample (TOC concentrations and the individual components making up the TOC fraction). In general, high-temperature (combustion) analyzers achieve more complete oxidation of the carbon fraction than low-temperature (wet chemistry/UV) analyzers. This can be important both in distinguishing different fractions of the organics in a sample and in achieving a precise measurement of the organic content of the sample.

The following are three different methods that are also available for detection and quantification of carbon dioxide produced in the oxidation step of a TOC analyzer:

- Nondispersive infrared detector
- Colorimetric methods
- Aqueous conductivity methods

The most common detector that on-line TOC analyzers use for source water and drinking water analysis is the nondispersive infrared detector.

While the differences in analytical methods employed by different TOC analyzers may be important in compliance or process monitoring, high levels of precision and the ability to distinguish specific organic fractions from a sample may not be required for detection of a potential chemical threat. Instead, gross deviations from normal TOC concentrations may be the best indication of a chemical threat to the system.

The detection limit for organic carbon depends on the measurement technique used (high or low temperature) and the type of analyzer. Because TOC concentrations are simply surrogates that can indicate potential problems in a system, gross changes in these concentrations are the best indicators of potential threats. Therefore, high-sensitivity probes may not be required for security purposes. However, the following detection limits can be expected:

- High temperature method (between 680°C and 950°C or higher in a few special cases, best possible oxidation): 1 mg/L carbon
- Low temperature method (below 100°C, limited oxidation potential): 0.2 mg/L carbon

The response time of a TOC analyzer may vary depending on the manufacturer's specifications, but it usually takes from five to fifteen minutes to get a stable, accurate reading.

Chemical Sensor—Chlorine Measurement System

Residual chlorine is one of the most sensitive and useful indicator parameters in water distribution system monitoring. All water distribution systems monitor for residual chlorine concentrations as part of their SDWA requirements, and procedures for monitoring chlorine concentrations are well established and accurate. Chlorine monitoring assures proper residual at all points in the system, helps pace rechlorination when needed, and quickly and reliably signals any unexpected increase in disinfectant demand. A significant decline or loss of residual chlorine could be an indication of potential threats to the system.

Several key points regarding residual chlorine monitoring for security purposes are provided below:

- Chlorine residuals can be measured using continuous on-line monitors at fixed points in the system or by taking grab samples at any point in the system and using chlorine test kits or portable sensors to determine chlorine concentrations.

- Correct placement of residual chlorine monitoring points within a system is crucial to early detection of potential threats. For example, while dead ends and low-pressure zones are common trouble spots that can show low residual chlorine concentrations, these zones are generally not of great concern for water security purposes because system hydraulics will limit the circulation of any contaminants present in these areas of the system.
- Monitoring procedures for SDWA compliance may be different from monitoring procedures for system security purposes, and utilities must determine the best use of on-line, fixed monitoring systems or portable sensors/test kits to balance their SDA compliance and security needs.

A variety of different portable and on-line chlorine monitors are commercially available. These range from sophisticated on-line chlorine monitoring systems to portable electrode sensors to colorimetric test kits. On-line systems can be equipped with control, signal, and alarm systems that notify the operator of low chlorine concentrations, and some may be tied into feedback loops that automatically adjust chlorine concentrations in the system. In contrast, use of portable sensors or colorimetric test kits requires technicians to take a sample and read the results. The technician then initiates required actions based on the results of the test.

The following are several measurement methods currently available to measure chlorine in water samples:

- N, N-diethyl-p-phenylenediamine (DPD) colorimetric method
- Iodometric method
- Amperometric electrodes
- Polarographic membrane sensors

The chlorine measurement devices here use the amperometric, DPD, or polarographic membrane sensor methods. It should be noted that there can be differences in the specific type of analyte, the range, and the accuracy of these different measurement methods. In addition, these different methods have different operations and maintenance requirements. For example, DPD systems require periodic replenishment of buffers, whereas polarographic systems do not. Users may want to consider these requirements when choosing the appropriate sensor for their systems.

Chemical Sensor—Portable Cyanide Analyzer

Portable cyanide detection systems are designed to be used in the field to evaluate for potential cyanide contamination of a water asset. These detection systems use one of two distinct analytical methods—either a colorimetric method or an ion selective method—to provide a quick, accurate cyanide measurement that does not require laboratory evaluation.

Aqueous cyanide chemistry can be complex. Various factors, including the water asset's pH and redox potential, can affect the toxicity of cyanide in that asset. While personnel using these cyanide detection devices do not need to have advanced knowledge of cyanide chemistry to successfully screen a water asset for cyanide, understanding aqueous cyanide chemistry can help users interpret whether the asset's cyanide concentration represents a potential threat. Therefore, a short summary of aqueous cyanide chemistry, including a discussion of cyanide toxicity, is provided blow. For more information, the reader is referred to *Standard Methods*, 20th Edition, Method 4500-CN.

Cyanide (CN-) is a toxic carbon-nitrogen organic compound that is the functional portion of the lethal gas hydrogen cyanide (HCN). The toxicity of aqueous cyanide varies depending on its form. At near-neutral pH, free cyanide (which is commonly designated as CN- although it is actually defined as the total of HCN and CN-) is the predominant cyanide form in water. Free cyanide is potentially toxic in its aqueous form, although the primary concern regarding aqueous cyanide is that it could volatilize. Free cyanide is not highly volatile (it is less volatile than most VOCs, but its volatility increases as the pH decreases below 8). However, when free cyanide does volatilize, it volatilizes in its highly toxic gaseous form (gaseous HCN). As a general rule, metal-cyanide complexes are much less toxic than free cyanide because they do not volatilize unless the pH is low.

Analyses for cyanide in public water systems are often conducted in certified labs using various EPA-approved methods, such as the preliminary distillation procedure with subsequent analysis by a colorimetric, ion-selective electrode, or flow injection methods. Lab analyses using these methods require careful sample preservation and pretreatment procedures and are generally expensive and time consuming. Using these methods, the following cyanide fractions are typically defined:

- **Total cyanide:** includes free cyanide (CN- + HCN) and all metal-completed cyanide.
- **Weak acid dissociable (WAD) cyanide:** includes free cyanide (CN- + HCN) and weak cyanide complexes that could be potentially toxic by hydrolysis to free cyanide in the pH range of 4.5–6.0.
- **Amendable cyanide:** includes free cyanide (CN- + HCN) and weak cyanide complexes that can release free cyanide at high pH (11–12). (This fraction gets its name because it includes measurement of cyanide from complexes that are "amendable" to oxidation by chlorine at high pH.) To measure amenable cyanide, the sample is split into two fractions. One of the fractions is analyzed for total cyanide as above. The other fraction is treated with high levels of chlorine for approximately one hour,

dechlorinated, and distilled per the above total cyanide method. Amendable cyanide is determined by the difference in the cyanide concentrations in these to fractions.

- **Soluble cyanide:** measures only soluble cyanide. Soluble cyanide is measured by using the preliminary filtration step, followed by the total cyanide analysis described above.

As discussed above, these different methods yield various cyanide measurements which may or may not give a complete picture of that sample's potential toxicity. For example, the total cyanide method includes cyanide complexed with metals, some of which will not contribute to cyanide toxicity unless the pH is out of the normal range. In contrast, the WAD cyanide measurement includes metal-complexed cyanide that could become free cyanide at low pH, and amendable cyanide measurements include metal-complexed cyanide that could become free cyanide at high pH. Personnel using these kits should therefore be aware of the potential differences in actual cyanide toxicity versus the cyanide measured in the sample under different environmental conditions.

Ingestion of aqueous cyanide can result in numerous adverse health effects and may be lethal. The EPA's MCL (an enforceable drinking water standard maintained by the EPA) for cyanide in drinking water is 0.2 mug/L (0.2 parts per million, or ppm). This MCL is based on free cyanide analysis per the amendable cyanide method described above. (The EPA has recognized that very stable metal-cyanide complexes such as iron-cyanide complex are nontoxic [unless exposed to significant UV radiation], and these fractions are therefore not considered when defining cyanide toxicity.) Ingestion of free cyanide at concentrations in excess of this MCL causes both acute effects (e.g., rapid breathing, tremors, and neurological symptoms) and chronic effects (e.g., weight loss, thyroid effects, and nerve damage). Under the current primary drinking water standards, public water systems are required to monitor their systems to minimize public exposure to cyanide levels in excess of the MCL.

HCN gas is also toxic, and the Office of Safety and Health administration has set a permissible exposure limit of 10 parts per million by volume (ppmv) for HCN inhalation. HCN also has a strong, bitter, almond-like smell and an odor threshold of approximately 1 ppmv. Considering the fact that HCN is relatively nonvolatile (see above), a slight cyanide odor emanating from a water sample suggests very high aqueous cyanide concentrations (greater than 10 to 50 mg/L, which is in the range of a lethal or near lethal dose with the ingestion of one pint of water).

Portable Field Monitors to Measure VOCs

VOCs are a group of highly utilized chemicals that have widespread applications, including use as fuel components, solvents, and cleaning and liquefying agents in degreasers, polishes, and dry cleaning solutions. VOCs are also used in herbicides and insecticides for agriculture applications.

Laboratory-based methods for analyzing VOCs are well established; however, analyzing VOCs in the lab is time consuming–obtaining a result may require several hours to several weeks depending on the specific method. Faster, commercially available methods for analyzing VOCs quickly in the field include use of portable gas chromatographs, mass spectrometers, or gas chromatographs/mass spectrometers, all of which can be used to obtain VOC concentration results within minutes. These instruments can be useful in rapid confirmation of the presence of VOCs in an asset or for monitoring an asset on a regular basis. In addition, portable VOC analyzers can analyze for a wide range of VOCs, such as toxic industrial chemicals, chemical warfare agents, drugs, explosives, and aromatic compounds.

There are several easy-to-use, portable VOC analyzers currently on the market that are effective in evaluating VOC concentrations in the field. These instruments utilize gas chromatography, mass spectroscopy, or a combination of both methods to provide near laboratory-quality analysis for VOCs.

Radiation Detection Equipment

Radioactive substances (radionuclides) are known health hazards that emit energetic waves and/or particles that can cause both carcinogenic and noncarcinogenic health effects. Radionuclides pose unique threats to source water supplies and water treatment, storage, or distribution systems because radiation emitted from radionuclides in water systems can affect individuals through several pathways—by direct contact with, ingestion or inhalation of, or external exposure to the contaminated water. While radiation can occur naturally in some cases due to the decay of some minerals, intentional and unintentional releases of man-made radionuclides into water systems is also a realistic threat.

Threats to water and wastewater facilities from radioactive contamination could involve two major scenarios. First, the facility or its assets could be contaminated, preventing workers from accessing and operating the facility/assets. Second, at drinking water facilities, the water supply could be contaminated, and tainted water could be distributed to users downstream. These two scenarios require different threat reduction strategies. The first scenario requires that facilities monitor for radioactive substances being brought onsite; the second requires that water assets be monitored for radioactive contamination. While the effects of radioactive contamination are basically the same under both threat types, each of these threats requires different types of radiation monitoring and different types of equipment.

Radiation Detection Equipment for Monitoring Water Assets

Most water systems are required to monitor for radioactivity and certain radionuclides and to meet Maximum MCLs for these contaminants to comply with the SDWA. Currently, the EPA requires drinking water to meet MCLs for beta/photon emitters (includes gamma radiation), alpha particles, combined radium 226/228, and uranium. However, this monitoring is required only at entry points into the system. In addition, after the initial sampling requirements, only one sample is required every three to nine years, depending on the contaminant type and the initial concentrations.

While this is adequate to monitor for long-term protection from overall radioactivity and specific radionuclides in drinking water, it may not be adequate to identify short-term spikes in radioactivity, such as from spills, accidents, or intentional releases. In addition, compliance with the SDWA requires analyzing water samples in a laboratory, which results in a delay in receiving results. In contrast, security monitoring is more effective when results can be obtained quickly in the field. In addition, monitoring for security purposes does not necessarily require that the specific radionuclides causing the contamination be identified. Thus, for security purposes, it may be more appropriate to monitor for non radionuclide-specific radiation using either portable field meters, which can be used as necessary to evaluate grab samples, or on-line systems, which can provide continuous monitoring of a system.

Ideally, measuring radioactivity in water assets in the field would involve minimal sampling and sample preparation. However, the physical properties of specific types of radiation combined with the physical properties of water make evaluating radioactivity in water assets in the field somewhat difficult. For example, alpha particles can only travel short distances and they cannot penetrate through most physical objects. Therefore, instruments designed to evaluate alpha emissions must be specially designed to capture emissions at a short distance from the source, and they must not block alpha emissions from entering the detector. Gamma radiation does not have the same types of physical properties, and thus it can be measured using different detectors.

Measuring different types of radiation is further complicated by the relationship between the radiation's intrinsic properties and the medium in which the radiation is being measured. For example, gas flow proportional counters are typically used to evaluate gross alpha and beta radiation from smooth, solid surfaces, but due to the fact that water is not a smooth surface, and because alpha and beta emissions are relatively short range and can be attenuated within the water, these types of counters are not appropriate for measuring alpha and beta activity in water. An appropriate method for measuring alpha and beta radiation in water is by using a liquid scintillation counter. However, this requires mixing an aliquot of water with a liquid scintillation "cocktail." The liquid scintillation counter is a large, sensitive piece of equipment, so it is not appropriate for field use. Therefore, measurements for alpha and beta radiation from water assets are not typically made in the field.

Unlike the problems associated with measuring alpha and beta activity in water in the field, the properties of gamma radiation allow it to be measured relatively well in water samples in the field. The standard instrumentation used to measure gamma radiation from water samples in the field is a sodium iodide scintillator.

Although the devices outlined above are the most commonly used for evaluating total alpha, beta, and gamma radiation, other methods and other devices can be used. In addition, local conditions (e.g., temperature, humidity) or the properties of the specific radionuclides emitting the radiation may make other devices or methods more optimal to achieve the goals of the survey than the devices noted above. Therefore, experts or individual vendors should be consulted to determine the appropriate measurement device for any specific application.

An additional factor to consider when developing a program to monitor for radioactive contamination in water assets is whether to take regular grab samples or sample continuously. For example, portable sensors can be used to analyze grab samples at any point in the system but have the disadvantage that they provide measurements only at one point in time. On the other hand, fixed-location sensors are usually used as part of a continuous, on-line monitoring system. These systems continuously monitor a water asset, and could be outfitted with some type of alarm system that would alert operators if radiation increased above a certain threshold. However, the sampling points are fixed and only certain points in the system can be monitored. In addition, the number of monitoring locations needed to capture the physical and radioactive complexity of a system can be prohibitive.

Toxicity Monitoring/Toxicity Meters

Toxicity measurement devices measure general toxicity to biological organisms, and detection of toxicity in any water/wastewater asset can indicate a potential threat, either to the treatment process (in the case of influent toxicity), to human health (in the case of drinking water toxicity), or to the environment (in the case of effluent toxicity). Currently, whole effluent toxicity tests (WET tests), in which effluent samples are tested against test organisms, are required of many NPDES discharge permits. The WET tests are used as a complement to the effluent limits on physical and chemical parameters to assess the overall effects of the discharge on living organisms or aquatic biota. Toxicity tests may also be used to monitor wastewater influent streams for potential hazardous contamination, such as organic heavy metals (arsenic, mercury, lead, chromium, and copper) that might upset the treatment process.

The ability to get feedback on sample toxicity from short-term toxicity tests or toxicity meters can be valuable in estimating the overall toxicity of a sample. On-line, real-time toxicity monitoring is still under active research and development. However, there are several portable toxicity measurement devices commercially available. They can generally be divided into the two categories below based on the different ways they measure toxicity:

- Meters measuring direct biological activity (e.g., luminescent bacteria) and correlating decreases in this direct biological activity with increased toxicity
- Meters measuring oxygen consumption and correlating decreases in oxygen consumption with increased toxicity.

COMMUNICATION/INTEGRATION

In this section, those devices necessary for communication and integration of water/wastewater system operations, such as electronic controllers, two-way radios, and wireless data communications, are discussed. Typically, SCADA systems would also be discussed in this section; however, SCADA is discussed in detail in chapter 7.

In regard to security applications, electronic controllers are used to automatically activate equipment (such as lights, surveillance cameras, audible alarms, or locks) when they are triggered. Triggering could be in response to a variety of scenarios, including tripping of an alarm or a motion sensor, breaking of a window or a glass door, variation in vibration sensor readings, or simply through input from a timer.

Two-way wireless radios allow two or more users that have their radios tuned to the same frequency to communicate instantaneously. They can communicate without the radios being physically lined together with wires or cables.

Wireless data communications devices are used to enable transmission of data between computer systems and/or between a SCADA server and its sensing devices, without individual components being physically linked together via wires or cables. In water and wastewater utilities, these devices are often used to link remote monitoring stations (i.e., SCADA components) or portable computers (i.e., laptops) to computer networks without using physical wiring connections.

Electronic Controllers

An electronic controller is a piece of electronic equipment that receives incoming electronic signals and uses preprogrammed logic to generate electronic output signals based on the incoming signals. While electronic controllers can be implemented for any application that involves inputs and outputs (e.g., control of a piece of machinery in a factory), in a security application, these controllers essentially act as the system's brain and can respond to specific security-related inputs with preprogrammed output responses. These systems combine the control of electronic circuitry with a logic function such that circuits are opened and closed (and thus equipment is turned on and off) through some preprogrammed logic. The basic principle behind the operation of an electrical controller is that it receives electronic inputs from sensors or any device generating an electrical signal (e.g., electrical signals from motion sensors), and then uses its preprogrammed logic to produce electrical outputs (e.g., these outputs could turn on power to a surveillance camera or to an audible alarm). Thus, these systems automatically generate a preprogrammed, logical response to a preprogrammed input scenario.

The three major types of electronic controllers are timers, electromechanical relays, and programmable logic controllers (PLCs), which are often called "digital relays." Each of these types of controller is discussed in more detail below.

Timers use internal signal/inputs (in contrast to externally generated inputs) and generate electronic output signals at certain times. More specifically, timers control electric current flow to any application to which they are connected, and can turn the current on or off on a schedule pre-specified by the user. Typical timer range (amount of time that can be programmed to elapse before the timer activates linked equipment) is from 0.2 seconds to 10 hours, although some of the more advanced timers have ranges of up to 60 hours. Timers are useful in fixed applications that don't require frequent schedule changes. For example, a timer can be used to turn on the lights in a room or building at a certain time every day. Timers are usually connected to their own power supply (usually 120-240 V).

In contrast to timers, which have internal triggers based on a regular schedule, electromechanical relays and PLCs have both external inputs and external outputs. However, PLCs are more flexible and more powerful than electromechanical relays, and thus this section focuses primarily on PLCs as the predominant technology for security-related electronic control applications.

Electromechanical relays are simple devices that use a magnetic field to control a switch. Voltage applied to the relay's input coil creates a magnetic field, which attracts an internal metal switch. This causes the relay's contacts to touch, closing the switch and completing the electrical circuit. This activates any linked equipment. These types of systems are often used for high-voltage applications, such as in some automotive and other manufacturing processes.

Two-Way Radios

Two way radios, as discussed here, are limited to a direct unit-to-unit radio communication, either via single unit-to-unit transmission and reception or via multiple hand-held units to a base station radio contact and distribution system. Radio frequency spectrum limitations apply to all hand-held units and are directed by the Federal Communications Commission. These limitations also distinguish a hand-held unit from a base station or base station unit (such as those used by an amateur [ham] radio operator), which operates under different wavelength parameters.

Two-way radios allow a user to contact another user or group of users instantly on the same frequency and to transmit voice or data without the need for wires. They use half-duplex communications, or communication that can be only transmitted or received; it cannot transmit and receive simultaneously. In other words, only one person may talk, while other personnel with radio(s) can only listen. To talk, the user depresses the talk button and speaks into the radio. The audio then transmits the voice wirelessly to the receiving radios. When the speaker has finished speaking and the channel has cleared, users on any of the receiving radios can transmit, either to answer the first transmission or to begin a new conversation. In addition to carrying voice data, many types of wireless radios also allow the transmission of digital data, and these radios may be interfaced with computer networks that can use or track these data. For example, some two-way radios can send global positioning system (GPS) data or the ID of the radio. Some two-way radios can also send data through a SCADA system.

Wireless radios broadcast these voice or data communications over the airwaves from the transmitter to the receiver. While this can be an advantage in that the signal emanates in all directions and does not need a direct, physical connection to be received at the receiver, it can also make the communications vulnerable to being blocked, intercepted, or otherwise altered. However, security features are available to ensure that the communications are not tampered with.

Wireless Data Communications

A wireless data communication system consists of two components: a wireless access point (WAP) and a wireless network interface card (sometimes also referred to as a client), which work together to complete the communications link. These wireless systems can link electronic devices, computers, and computer systems together using radio waves, thus eliminating the need for these individual components to be directly connected through physical wires. While wireless data communications have widespread applications in water and wastewater systems, they also have limitations. First, wireless data connections are limited by the distance between components (radio waves scatter over a long distance and cannot be received efficiently unless special directional antennae are used). Second, these devices only function if the individual components are in a direct line of sight with each other, since radio waves are affected by interference from physical obstructions. However, in some cases, repeater units can be used to amplify and retransmit wireless signals to circumvent these problems. The two components of wireless devices are discussed in more detail below.

The WAP provides the wireless data communication service. It usually consists of a housing (which is constructed from plastic or metal depending on the environment it will be used in) containing a circuit board, flash memory that holds software, one or two external ports to connect to existing wired networks, a wireless radio transmitter/receiver, and one or more antenna connections. Typically, the WAP requires a one-time user configuration to allow the device to interact with the local area network (LAN). This configuration is usually done via a web-driven software application, which is accessed with a computer.

A wireless network interface card/client is a piece of hardware that is plugged into a computer and enables that computer to make a wireless network connection. The card consists of a transmitter, functional circuitry, and a receiver for the wireless signal, all of which work together to enable communication between the computer, its wireless transmitter/receiver, and its antenna connection. Wireless cards are installed in a computer through a variety of connections, including USB adapters or Laptop CardBus or desktop peripheral cards. As with the WAP, software is loaded onto the user's computer, allowing configuration of the card so that it may operate over the wireless network.

Two of the primary applications for wireless data communications systems are to enable mobile or remote connections to a LAN, and to establish wireless communications links between SCADA remote telemetry units (RTUs) and sensors in the field. Wireless car connections are usually used for LAN access from mobile computers. Wireless cards can also be incorporated into RTUs to allow them to communicate with sensing devices that are located remotely.

CYBER PROTECTION DEVICES

Various cyber protection devices are currently available for use in protecting utility computer systems. These protection devices include antivirus and pest eradication software, firewalls, and network intrusion hardware/software.

Anti-virus and Pest Eradication Software

Antivirus programs are designed to detect, delay, and respond to programs or pieces of code that are specifically designed to harm computers. These programs are known as "malware." Malware can include computer viruses, worms, and Trojan Horse programs (programs that appear to be benign but have hidden harmful effects).

Pest eradication tools are designed to detect, delay, and respond to spyware (strategies that websites use to track user behavior, such as by sending "cookies" to the user's computer), and hacker tools that track keystrokes (keystroke loggers) or passwords (password crackers).

Viruses and pests can enter a computer system through the Internet or through infected floppy discs or CDs. They can also be placed onto a system by insiders. Some of these programs, such as viruses and worms, move within a computer's drives and files or between computers if the computers are networked to each other. This malware can deliberately damage files, utilize memory and network capacity, crash application programs, and initiate transmissions of sensitive information from a PC. While the specific mechanisms of these programs differ, they can infect files, and even the basic operating program of the computer firmware/hardware.

The most important features of an antivirus program are its abilities to identify potential malware and to alert a user before infection occurs, as well as its ability to respond to a virus already resident on a system. Most of these programs provide a log so that the user can see what viruses have been detected and where they were detected. After detecting a virus, the antivirus software may delete the virus automatically, or it may prompt the user to delete the virus. Some programs will also fix files or programs damaged by the virus.

Various sources of information are available to inform the general public and computer system operators about new viruses being detected. Since antivirus programs use signatures (snippets of code or data) to detect the presence of a virus, periodic updates are required to identify new threats. Many antivirus software providers offer free upgrades that detect and respond to the latest viruses.

Firewalls

A firewall is an electronic barrier designed to keep computer hackers, intruders, or insiders from accessing specific data files and information on a utility's computer network or other electronic/computer systems. Firewalls operate by evaluating and then filtering information coming through a public network (such as the Internet) into the utility's computer or other electronic system. This evaluation can include identifying the source or destination addresses and ports and allowing or denying access based on this identification.

The following are two methods used by firewalls to limit access to the utility's computers or other electronic systems from the pubic network:

- The firewall may deny all traffic unless it meets certain criteria.
- The firewall may allow all traffic through unless it meets certain criteria.

A simple example of the first method is to screen requests to ensure that they come from an acceptable (e.g., previously identified) domain name and Internet protocol address. Firewalls may also use more complex rules that analyze the application data to determine if the traffic should be allowed through. For example, the firewall may require user authentication (e.g., use of a password) to access the system. How a firewall determines what traffic to let through depends on which network layer it operates at and how it is configured. Some of the pros and cons of various methods to control traffic flowing in and out of the network are provided in Table 9.12.

Firewalls may be a piece of hardware, a software program, or an appliance card that contains both.

Advanced features that can be incorporated into firewalls allow for the tracking of attempts to log on to the local area network system. For example, a report of successful and unsuccessful log-on attempts may be generated for the computer specialist to analyze. For systems with mobile users, firewalls allow remote access to the private net-

Table 9.12. Pros and Cons of Various Firewall Methods for Controlling Network Access

Method	*Description*	*Pros*	*Cons*
Packet filtering	Incoming and outgoing packets (small chunks of data) are analyzed against a set of filters. Packets that make it through the filters are sent to the requesting system and all others are discarded. There are two type of packet filtering: static (the most common) and dynamic	Static filtering is relatively inexpensive, and little maintenance is required; well -suited for closed environments where access to or from multiple addresses is not allowed	Leaves permanent open holes in the network; allows direct connection to internal hosts by external sources; offers no user authentication; can be unmanageable in large networks
Proxy service	Information from the Internet is retrieved by the firewall and then sent to the requesting system and vice versa. In this way, the firewall can limit the information made known to the requesting system, making vulnerabilities less apparent	Only allows temporary open holes in the network perimeter; can be externally used for all types of internal protocol services	Allows direct connections from internal host to clients; offers no user authentication
Stateful pattern recognition	This method examines and compares the contents of certain key parts of an information packet against a database of acceptable information. Information traveling from inside the firewall to the outside is monitored for specific defining characteristics; then, incoming information is compared to these characteristics. If the comparison yields a reasonable match, the information is allowed through. If not, the information is discarded.	Provides a limited time window to allow pockets of information to be sent; does not allow any direct connections between internal and external hosts; supports user-level authentication	Slower than packet filtering; does not support all types of connections

Source: Water and Wastewater Security Product Guide, U.S. EPA, 2005.

work by the use of secure log-on procedures and authentication certificates. Most firewalls have a graphical user interface for managing the firewall.

In addition, new Ethernet firewall cards that fit in the slot of an individual computer bundle additional layers of defense (like encryption and permit/deny) for individual computer transmissions to the network interface function. These new cards have only a slightly higher cost than traditional network interface cards.

Network Intrusion Hardware/Software

Network intrusion detection and prevention systems are software- and hardware-based programs designed to detect unauthorized attacks on a computer network system.

While other applications, such as firewalls and antivirus software, share similar objectives with network intrusion systems, network intrusion systems provide a deeper layer of protection beyond the capabilities of these other systems because they evaluate patterns of computer activity rather than specific files.

It is worth noting that attacks may come from either outside or within the system (i.e., from an insider) and that network intrusion detection systems may be more applicable for detecting patterns of suspicious activity from inside a facility (e.g., accessing sensitive data) than other information technology solutions.

Network intrusion detection systems employ a variety of mechanisms to evaluate potential threats. The types of search and detection mechanisms are dependent on the level of sophistication of the system. Some of the available detection methods include the following:

- **Protocol analysis:** Protocol analysis is the process of capturing, decoding, and interpreting electronic traffic. The protocol analysis method of network intrusion detection involves the analysis of data captured during transactions between two or more systems or devices, and the evaluation of these data to identify unusual activity and potential problems. Once a problem is isolated and recorded, problems or potential threats can be linked to pieces of hardware or software. Sophisticated protocol analysis will also provide statistics and trend information on the captured traffic.
- **Traffic anomaly detection:** Traffic anomaly detection identifies potential threatening activity by comparing incoming traffic to normal traffic patterns and identifying deviations. It does this by comparing user characteristics against thresholds and triggers defined by the network administrator. This method is designed to detect attacks that span a number of connections rather than a single session.
- **Network honeypot:** This method establishes nonexistent services in order to identify potential hackers. A network honeypot impersonates services that don't exist by sending fake information to people scanning the network. It identifies the attacker

when they attempt to connect to the service. There is no reason for legitimate traffic to access these resources because they don't exist; therefore any attempt to access them constitutes an attack.

- **Anti-intrusion detection system evasion techniques:** These methods are designed for attackers who may be trying to evade intrusion detection system scanning. They include methods called IP defragmentation, TCP streams reassembly, and deobfuscation.

While these detection systems are automated, they can only indicate patterns of activity, and a computer administrator or other experienced individual must interpret activities to determine whether or not they are potentially harmful. Monitoring the logs generated by these systems can be time consuming, and there may be a learning curve to determine a baseline of normal traffic patterns from which to distinguish potential suspicious activity.

REFERENCES

Garcia, M. L. 2001. *The Design and Evaluation of Physical Protection Systems.* London: Butterworth-Heinemann.

IBWA. 2004. *Bottled Water Safety and Security.* Alexandria, VA: International Bottled Water Association.

Water and Wastewater Security Product Guide. 2005. U.S. EPA. http://cfpub.epa.gov.safewater/watersecurity/guide.

10

A Changed Nation

By turns intimate, indifferent, and intemperate; post-9/11, a target. Our waters.

—*F. R. Spellman*

The events of 9/11 dramatically changed this nation and focused us on combating terrorism. As a result, in fall 2003, the National Drinking Water Advisor Council (NDWAC)—composed of members from the general public, state and local agencies, and private groups concerned with safe drinking water, who advise the Environmental Protection Agency (EPA) administrator on everything that the agency does relating to drinking water—established a Water Security Working Group (WSWG) to consider and make recommendations on water security issues. The NDWAC directed the WSWG to do the following:

- Identify active and effective security practices for drinking water and wastewater utilities, and provide an approach for adopting these practices.
- Recommend mechanisms to provide incentives that facilitate broad and receptive response among the water sector to implement active and effective security practices.
- Recommend mechanisms to measure progress and achievements in implementing active and effective security practices, and identify barriers to implementation.

The NDWAC recommendations on security are structured to maximize benefits to utilities by emphasizing actions that have the potential both to improve the quality or reliability of utility service and to enhance security. The recommendations were designed for use by water systems of all types and sizes, including systems that serve less than 3,300 people.

The NDWAC identified 14 features of active and effective security programs that are important to increasing security and relevant across the broad range of utility circumstances and operating conditions. The EPA (2003) points out that the 14 features are, in many cases, consistent with the steps needed to maintain technical, management, and operational performance capacity related to overall water quality. Many utilities may be able to adopt some of the features with minimal, if any, capital investment.

FOURTEEN FEATURES OF ACTIVE AND EFFECTIVE SECURITY

It is important to point out that the 14 features of active and effective programs emphasize that one size does not fit all, and that there will be variability in security approaches and tactics among water utilities based on utility-specific circumstances and operating conditions. The following are characteristics of the 14 features:

- They are sufficiently flexible to apply to all utilities, regardless of size.
- They incorporate the idea that active and effective security programs should have measurable goals and time lines.
- They allow flexibility for utilities to develop specific security approaches and tactics that are appropriate to utility-specific circumstances.

Water utilities can differ in many ways, including the following:

- Supply source (groundwater, surface water, etc.)
- Number of supply sources
- Treatment capacity
- Operation risk
- Location risk
- Security budget
- Spending priorities
- Political and public support
- Legal barriers
- Public vs. private ownership

The EPA (2003) recommends that all utilities address security in an informed and systematic way, regardless of these differences. Utilities need to fully understand the specific local circumstances and conditions under which they operate and develop a security program tailored to those conditions. The goal in identifying common features of active and effective security programs is to achieve consistency in security program outcomes among water utilities, while allowing for and encouraging utilities to develop utility-specific security approaches and tactics. The features are based on a comprehensive security management layering system approach that incorporates a combination of public involvement and awareness; partnerships; and physical, chemical, operational, and design controls to increase overall program performance. They address utility security in four functional categories: organization, operation, infrastructure, and external. These functional categories are discussed in greater detail below.

- **Organizational:** There is always something that can be done to improve security. Even when resources are limited, the simple act of increasing organizational attentiveness to security may reduce vulnerability and increase responsiveness. Preparedness itself can help deter attacks. The first step to achieving preparedness is to make security a part of the organizational culture, so that it is in the day-to-day thinking of frontline employees, emergency responders, and management of every water and wastewater utility in this country. To successfully incorporate security into "business as usual," there must be a strong commitment to security by organization leadership and by the supervising body, such as the utility board or rate-setting organization. The following features address how a security culture can be incorporated into an organization.
- **Operational:** In addition to having a strong culture and awareness of security within an organization, an active and effective security program makes security part of operational activities, from daily operations—such as monitoring physical access controls—to scheduled annual reassessments. Utilities will often find that by implementing security into operations, they can also reap cost benefits and improve the quality or reliability of utility service.
- **Infrastructure:** These recommendations advise utilities to address security in all elements of utility infrastructure—from source water to distribution and through collection and wastewater treatment.
- **External:** Strong relationships with response partners and the public strengthen security and public confidence. Two of the recommended features of active end effective security programs address this need.

Fourteen Features

Feature 1. Senior leadership must make an explicit and visible commitment to security.

Utilities should create an explicit, easily communicated, enterprise-wide commitment to security, which can be done through the following:

- Incorporate security into a utility-wide mission or vision statement that addresses the full scope of an active and effective security program; the vision statement should include protection of public health, public safety, and public confidence as part of core, day-to-day operations.
- Develop an enterprise-wide security policy or set of policies.

Utilities should use the process of making a commitment to security as an opportunity to raise awareness of security throughout the organization, making the commitment visible to all employees and customers and helping every facet of the enterprise recognize the contribution they can make to enhancing security.

Feature 2. Promote security awareness throughout the organization.

The objective of a security culture should be to make security awareness a normal, accepted, and routine part of day-to-day operations. Examples of tangible efforts include the following:

- Conducting employee training
- Incorporating security into job descriptions
- Establishing performance standards and evaluations for security
- Creating and maintaining a security tip line and suggestion box for employees
- Making security a routine part of staff meetings and organization planning
- Creating a security policy

Feature 3. Assess vulnerabilities and periodically review and update vulnerability assessments to reflect changes in potential threats and vulnerabilities.

Because circumstances change, utilities should maintain their understanding and assessment of vulnerabilities as a living document and continually adjust their security enhancement and maintenance priorities. Utilities should consider their individual circumstances and establish and implement a schedule for review of their vulnerabilities.

Assessments should take place once every three to five years at a minimum. Utilities may be well served by doing assessments annually.

The basic elements of sound vulnerability assessments include the following:

- Characterization of the water system, including its mission and objectives
- Identification and prioritization of adverse consequences to avoid
- Determination of critical assets that might be subject to malevolent acts
- Assessment of the likelihood (qualitative probability) of such malevolent acts from adversaries
- Evaluation of existing countermeasures
- Analysis of current risk and development of a prioritized plan for risk reduction

Feature 4. Identify security priorities and, on an annual basis, identify the resources dedicated to security programs and planned security improvements, if any.

Dedicated resources are important to ensure a sustained focus on security. Investment in security should be reasonable considering utilities' specific circumstances. In some circumstances, investment may be as simple as increasing the amount of time and attention that executives and managers give to security. Where threat potential or potential consequences are greater, greater investment is likely warranted.

This feature establishes the expectation that utilities should, through their annual capital, operations, maintenance, and staff resources plan, identify and set aside resources consistent with their specific identified security needs. Security priorities should be clearly documented and should be reviewed with utility executives at least once a year as part of the traditional budgeting process.

Feature 5. Identify managers and employees who are responsible for security and establish security expectations for all staff.

Consider the following when addressing this feature:

- Explicit identification of security responsibilities is important for development of a security culture with accountability.
- At minimum, utilities should identify a single designated individual responsible for overall security, even if other security roles and responsibilities will likely be dispersed throughout the organization.
- The number and depth of security-related roles will depend on a utility's specific circumstances.

Feature 6. Establish physical and procedural controls to restrict access to utility infrastructure to those conducting authorized, official business and to detect unauthorized physical intrusions.

Examples of physical access controls include fencing critical areas, locking gates and doors, and installing barriers at site access points. Monitoring for physical intrusion can include maintaining well-lighted facility perimeters, installing motion detectors, and utilizing intrusion alarms. The use of neighborhood watches, regular employee rounds, and arrangements with local police and fire departments can support identifying unusual activity in the vicinity of facilities.

Examples of procedural access controls include inventorying keys, changing access codes regularly, and requiring security passes to pass gates an access-sensitive area. In addition, utilities should establish the means to readily identify all employees including contractors and temporary workers with unescorted access to facilities.

Feature 7. Establish employee protocols for detection of contamination consistent with the recognized limitations in current contaminant detection, monitoring, and surveillance technology.

Until progress can be made in development of practical and affordable on-line contaminant monitoring and surveillance systems, most utilities must use other approaches to contaminant monitoring and surveillance. This includes monitoring data of physical and chemical contamination surrogates, pressure change abnormalities, free and total chlorine residual, temperature, dissolved oxygen, and conductivity.

Many utilities already measure the above parameters on a regular basis to control plant operations and confirm water quality; more closely monitoring these parameters may create operational benefits for utilities that extend far beyond security, such as reducing operating costs and chemical usage. Utilities should also thoughtfully monitor customer complaints and improve connections with local public health networks to detect public health anomalies. Customer complaints and public health anomalies are an important way to detect potential contamination problems and other water quality concerns.

Feature 8. Define security-sensitive information; establish physical, electronic, and procedural controls to restrict access to security-sensitive information; detect unauthorized access; and ensure information and communications systems will function during emergency response and recovery.

Protecting IT systems largely involves using physical hardening and procedural steps to limit the number of individuals with authorized access and to prevent access by unauthorized individuals. Examples of physical steps to harden SCADA and IT networks include installing and maintaining fire walls and screening the network for viruses. Examples of procedural steps include restricting remote access to data networks and safeguarding critical data through backups and storage in safe places. Utilities should strive for continuous operation of IT and telecommunications systems, even in the event of an attack, by providing uninterruptible power supply and back up systems, such as satellite phones.

In addition to protecting IT systems, security-sensitive information should be identified and restricted to the appropriate personnel. Security sensitive information could be contained within the following:

- Facility maps and blueprints
- Operations details
- Hazardous material utilization
- Tactical level security program details
- Any other information on utility operations or technical details that could aid in planning or execution of an attack

When identifying security-sensitive information, consider all ways that utilities might use and make public information (e.g., many utilities may at times engage in competitive bidding processes for construction of new facilities or infrastructure). Finally, information critical to the continuity of day-to-day operations should be identified and backed up.

Feature 9. Incorporate security considerations into decisions about acquisition, repair, major maintenance, and replacement of physical infrastructure; include considerations of opportunities to reduce risk through physical hardening and adoption of inherently lower-risk design and technology options.

Prevention is a key aspect of enhancing security. Consequently, consideration of security issues should begin as early as possible in facility construction (i.e., it should be a factor in building plans and designs). However, to incorporate security considerations into design choices, utilities need information about the types of security design approaches and equipment that are available and the performance of these designs and equipment in multiple dimensions. For example, utilities would want to not just evaluate the way that a particular design might contribute to security, but also look at how that design would affect the efficiency of day-to-day plant operations and worker safety.

Feature 10. Monitor available threat-level information and escalate security procedures in response to relevant threats.

Monitoring threat information should be a regular part of a security program manager's job, and utility-, facility-, and region-specific threat levels and information should be shared with those responsible for security. As part of security planning, utilities should develop systems to assess threat information—procedures that will be followed in the event of increased industry or facility threat levels—and should be prepared to put these procedures in place immediately so that adjustments are seamless. Involving local law enforcement and FBI is critical.

Utilities should investigate what networks and information sources might be available to them locally and at the state and regional level. If a utility cannot gain access to some information networks, attempts should be made to align with those who can and will provide effective information to the utility.

Feature 11. Incorporate security considerations into emergency response and recovery plans; test and review plans regularly; and update plans to reflect changes in potential threats, physical infrastructure, utility operations, critical interdependencies, and response protocols in partner organizations.

Utilities should maintain response and recovery plans as living documents. In incorporating security considerations into their emergency response and recovery plans, utilities also should be aware of the National Incident Management System (NIMS) guidelines, established by the Department of Homeland Security (DHS), and of regional and local incident management commands and systems, which tend to flow from the national guidelines. Adoption of NIMS guidelines is required to qualify for funds dispersed through the EPA and DHS.

Utilities should consider their individual circumstances and establish, develop, and implement a schedule for review of emergency response and recovery plans. Utility plans should be thoroughly coordinated with emergency response and recovery planning in the larger community. As part of this coordination, a mutual aid program should be established to arrange in advance for exchanging resources (personnel or physical assets) among agencies within a region in the event of an emergency or disaster that disrupts operation. Typically, the exchange of resource is based on a written formal mutual aid agreement. For example, Florida's Water-Wastewater Agency Response Network, deployed after Hurricane Katrina, allowed the new "utilities helping utilities" network to respond to urgent requests from Mississippi for help to bring facilities back online after the hurricane.

The emergency response and recovery plans should be reviewed and updated as needed annually. This feature also establishes the expectation that utilities should test or exercise their emergency response and recovery plans regularly.

Feature 12. Develop and implement strategies for regular, ongoing, security-related communications with employees, response organizations, rate-setting organizations, and customers.

An active and effective security program should address protection of public health, public safety (including infrastructure), and public confidence. Utilities should create an awareness of security and an understanding of the rationale for their overall security management approach in the communities they serve, including rate-setting organizations.

Effective communication strategies consider key messages; who is best equipped/trusted to deliver the key messages; the need for message consistency, particularly during an emergency; and the best mechanisms for delivering messages and for receiving information and feedback from key partners. The key audiences for communication strategies are utility employees, response organizations, and customers.

Feature 13. Forge reliable and collaborative partnerships with the communities served, managers of critical interdependent infrastructure, response organizations, and other local utilities.

Effective partnerships build collaborative working relationships and clearly define roles and responsibilities so that people can work together seamlessly if an emergency should occur. It is important for utilities within a region and neighboring regions to collaborate and establish a mutual aid program with neighboring utilities, response organizations, and sectors, such as the power sector, on which utilities rely or impact. Mutual aid agreements provide for help from other organizations that is prearranged and can be accessed quickly and efficiently in the event of a terrorist attack or natural disaster. Developing reliable and collaborative partnerships involves reaching out to managers, key staff, and other organizations to build reciprocal understanding and to share information about the utility's security concerns and planning. Such efforts will maximize the efficiency and effectiveness of a mutual aid program during an emergency response effort, as the organizations will be familiar with each others' circumstances, and thus will be better able to serve each other.

It is also important for utilities to develop partnerships with the communities and customers they serve. Partnerships help to build credibility within communities and establish public confidence in utility operations. People who live near utility structures ("water watchers") can be the eyes and ears of the utility and can be encouraged to notice and report changes in operating procedures or other suspicious behaviors.

Utilities and public health organizations should establish formal agreements on coordination to ensure regular exchange of information between utilities and public health organizations and should outline roles and responsibilities during response to and recovery from an emergency. Coordination is important at all levels of the public health community—national public health, county health agencies, and healthcare providers, such as hospitals.

Feature 14. Develop utility-specific measures of security activities and achievements, and self-assess against these measures to understand and document program progress.

Although security approaches and tactics will be different depending on utility-specific circumstances and operating conditions, we recommend that all utilities monitor and measure a number of common types of activities and achievements, including existence of program policies and procedures, training, testing, and implementing schedules and plans.

The 14 Feature Matrix

In table 10.1, a matrix of recommended measures to assess effectiveness of a utility's security program is presented. Each feature is grouped according to its functional category: organization, operation, infrastructure, or external.

Final Word on Security

Ultimately, the goal of implementing the 14 security features (and all other security provisions) is to create a significant improvement in water security on a national scale by reducing vulnerabilities and therefore risk to public health from terrorist attacks and natural disasters. To create a sustainable effect, the water/wastewater sector as a whole must not only adopt and actively practice the features, but also incorporate the features into business as usual.

REFERENCES

Active and Effective Water Security Programs. 2003. U.S. EPA, at cfpub.epa.gov/safewater/watersecurity/index.cfm.

Table 10.1. Fourteen Features of an Active and Effective Security Matrix

Features Checklist	*Potential Measures of Progress*
	Organizational Features
Feature 1: Explicit commitment to security	Does a written, enterprise-wide security policy exist, and is the policy reviewed regularly and updated as needed?
Feature 2: Promotion of security awareness	Are incidents reported in a timely way, and are lessons learned from incident responses reviewed and, as appropriate, incorporated into future utility security efforts?
Feature 5: Defined security roles and employee expectations	Are managers and employees who are responsible for security identified?
	Operational Features
Feature 3: Up-to-date vulnerability assessment	Are reassessments of vulnerabilities made after incidents, and are lessons learned and other relevant information incorporated into security practices?
Feature 4: Security resources and implementation priorities	Are security priorities clearly identified, and to what extent do security priorities have resources assigned to them?
Feature 7: Contamination detection	Is there a protocol/procedure in place to identify and respond to suspected contamination events?
Feature 10: Threat level-based protocols	Is there a protocol/procedure of responses that will be made if threat levels change?
Feature 11: Tested and up-to-date Emergency Response Plan	Do exercises address the full range of threats—physical, cyber, and contamination—and is there a protocol/procedure to incorporate lessons learned from exercises and actual responses into updates to emergency response and recovery plans?
Feature 14: Utility-specific measures and self-assessment	Does the utility perform self-assessment at least annually?
Infrastructure Features	
Feature 6: Intrusion detection and access control	To what extent are methods to control access to sensitive assets in place?
Feature 8: Information protection and continuity	Is there a procedure to identify and control security-sensitive information, is information correctly categorized, and how do control measures perform under testing?
Feature 9: Design and construction standards	Are security considerations incorporated into internal utility design and construction standards for new facilities/infrastructure and major maintenance projects?
External Features	
Feature 12: Communications	Is there a mechanism for utility employees, partners, and the community to notify the utility of suspicious occurrences and other security concerns?
Feature 13: Partnerships	Have reliable and collaborative partnerships with customers, managers of independent interrelated infrastructure, and response organizations been established?

Source: Active and Effective Water Security Programs, U.S. EPA, 2003.

Appendix

Safety Requirements of Sections 1433, 1434, and 1435 of the Safe Drinking Water Act

SECTION 1433. TERRORIST AND OTHER INTENTIONAL ACTS

(a) Vulnerability Assessments—

(1) Each community water system serving a population of greater than 3,300 persons shall conduct an assessment of the vulnerability of its system to a terrorist attack or other intentional acts intended to substantially disrupt the ability of the system to provide a safe and reliable supply of drinking water. The vulnerability assessment shall include, but not be limited to, a review of pipes and constructed conveyances, physical barriers, water collection, pretreatment, treatment, storage and distribution facilities, electronic, computer or other automated systems which are utilized by the public water system, the use, storage, or handling of various chemicals, and the operation and maintenance of such system. The Administrator, not later than August 1, 2002, after consultation with appropriate departments and agencies of the federal government and with state and local governments, shall provide baseline information to community water systems required to conduct vulnerability assessments regarding which kinds of terrorist attacks or other international acts are the probable threats to—

(A) substantially disrupt the ability of the system to provide a safe and reliable supply of drinking water;

or

(B) otherwise present significant public health concerns.

(2) Each community water system referred to in paragraph (1) shall certify to the Administrator that the system has conducted an assessment complying with paragraph (1) and shall submit to the Administrator a written copy of the assessment.

Such certification and submission shall be made prior to:

(A) March 31, 2003, in the case of systems serving a population of 100,000 or more.

(B) December 31, 2003, in the case of systems serving a population of 50,000 or more but less than 100,000.

(C) June 30, 2004, in the case of systems serving a population greater than 3,300 but less than 50,000.

(3) Except for information contained in a certification under this subsection identifying the system submitting the certification and the date of the certification, all information provided to the Administrator under this subsection and all information derived there from shall be exempt from disclosure under section 552 of title 5 of the United States Code.

(4) No community water system shall be required under state or local law to provide an assessment described in this section to any state, regional, or local governmental entity solely by reason of the requirement set forth in paragraph (2) that the system submit such assessment to the Administrator.

(5) Not later than November 30, 2002, the Administrator, in consultation with appropriate federal law enforcement and intelligence officials, shall develop such protocols as may be necessary to protect the copies of the assessments required to be submitted under this subsection (and the information contained therein) from unauthorized disclosure. Such protocols shall ensure that—

(A) each copy of such assessment, and all information contained in or derived from the assessment, is kept in a secure location;

(B) only individuals designated by the Administrator may have access to the copies of the assessments; and

(C) no copy of an assessment, or part of an assessment, or information contained in or derived from an assessment shall be available to anyone other than an individual designated by the Administrator.

At the earliest possible time prior to November 30, 2002, the Administrator shall complete the development of such protocols for the purpose of having them in place prior to receiving any vulnerability assessments for community water systems under this subsection.

(6) (A) Except as provided in subparagraph (B), any individual referred to in paragraph (5) (B) who acquires the assessment submitted under paragraph (2), or any reproduction of such assessment, or who knowingly or recklessly reveals such assessments, reproduction, or information other than—

(i) to an individual designated by the Administrator under paragraph (5),

(ii) for purposes of section 1445 or for actions under section 1431, or

(iii) for use in any administrative or judicial proceeding to impose a penalty for failure to comply with this section shall upon conviction be imprisoned for not more than one year or fined in accordance with the provisions of chapter 227 of title 18, United States Code, applicable to class A misdemeanors, or both, and shall be removed from federal office or employment.

(B) Notwithstanding subparagraph (A), an individual referred to in paragraph (5)(B) who is an officer or employee of the United States may discuss the contents of a vulnerability assessment submitted under this section with a state or local official.

(7) Nothing in this section authorizes any person to withhold any information from Congress or from any committee or subcommittee of Congress.

(b) Emergency Response Plan—

Each community water system serving a population greater than 3,300 shall prepare or revise, where necessary, an emergency response plan that incorporates the results of vulnerability assessments that have been completed. Each such community water system shall certify to the Administrator, as soon as is reasonably possible after the enactment of this section, but not later than 6 months after the completion of the vulnerability assessment under subsection (a), that the system has completed such plan. The emergency response plan shall include, but not be limited to, plans, procedures, and identification of equipment that can be implemented or utilized in the event of a terrorist or other intentional attack on the public water system. The emergency response shall also include actions, procedures, and identification of equipment which can obviate or significantly lessen the impact of terrorist attacks or other intentional actions on the public health and the safety and supply of drinking water provided to communities and individuals. Community water systems shall, to the extent possible, coordinate with existing Local Emergency Planning Committees established under the Emergency Planning and Community Right-to-Know Act (42 U.S.C. 11001 et seq.) when preparing or revising an emergency response plan under this subsection.

(c) Record Maintenance—

Each community water system shall maintain a copy of the emergency response plan completed pursuant to subsection (b) for 5 years after such plan has been certified to the Administrator under this section.

(d) Guidance to Small Public Water Systems—

The Administrator shall provide guidance to community water systems serving a population of less than 3,300 persons on how to conduct vulnerability assessments, prepare emergency response plans, and address threats from terrorist attacks or other intentional actions designed to disrupt the provision of safe drinking water or significantly affect the public health or significantly affect the safety or supply of drinking water provided to communities and individuals.

(e) Funding—

(1) There are authorized to be appropriated to carry out this section not more than $160,000,000 for the fiscal year 2002 and such sums as may be necessary for the fiscal years 2003 through 2005.

(2) The Administrator, in coordination with state and local governments, may use funds made available under paragraph (1) to provide financial assistance to community water systems for purposes of compliance with the requirements of subsections (a) and (b) and to community water systems for expenses and contracts designed to address basic security enhancements of critical importance and significant threats to public health and the supply of drinking water as determined by a vulnerability assessment conducted under subsection (a). Such basic security enhancements may include, but shall not be limited to the following:

(A) the purchase and installation of equipment for detection of intruders;

(B) the purchase and installation of fencing, gating, lighting, or security cameras;

(C) the tamper-proofing of manhole covers, fire hydrants, and valve boxes;

(D) the rekeying of doors and locks;

(E) improvements to electronic, computer, or to other automated systems and remote security systems;

(F) participation in training programs, and the purchase of training manuals and guidance materials, relating to security against terrorist attacks;

(G) improvements in the use, storage, or handling of various chemicals; and

(H) security screening of employees or contractor support services. Funding under this subsection for basic security enhancements shall not include expenditures for personnel costs, or monitoring, operation, or maintenance of facilities, equipment, or systems.

(3) The Administrator may use not more than $5,000,000 from the funds made available under paragraph (1) to make grants to community water systems to assist in responding to and alleviating any vulnerability to a terrorist attack or other intentional acts intended to substantially disrupt the ability of the system to provide a safe and reliable supply of drinking water (including sources of water for such systems) which the Administrator determines to present an immediate and urgent security need.

(4) The Administrator may use not more than $5,000,000 from the funds made available under paragraph (1) to make grants to community water systems serv-

ing a population of less than 3,300 persons for activities and projects undertaken in accordance with the guidance provided to such systems under subsection (d).

SECTION 1434. CONTAMINANT PREVENTION, DETECTION, AND RESPONSE

(a) In General—

The Administrator, in consultation with the Centers for Disease Control and, after consultation with appropriate departments and agencies of the federal government, and with state and local governments, shall review (or enter into contracts or cooperative agreements to provide for a review of) current and future methods to prevent, detect, and respond to the intentional introduction of chemical, biological, or radiological contaminants into community water systems and source water for community water systems, including each of the following:

(1) Methods, means, and equipment, including real-time monitoring systems, designed to monitor and detect various levels of chemical, biological, and radiological contaminants or indicators of contaminants and reduce the likelihood that such contaminants can be successfully introduced into public water systems and source water intended to be used for drinking water.
(2) Methods and means to provide sufficient notice to operators of public water systems, and individuals served by such systems, or the introduction of chemical, biological, or radiological contaminants and the possible effect of such introduction on public health and the safety and supply of drinking water.
(3) Methods and means for developing educational and awareness programs for community water systems.
(4) Procedures and equipment necessary to prevent the flow of contaminated drinking water to individuals served by public water systems.
(5) Methods, means, and equipment which could negate or mitigate deleterious effects on public health and the safety and supply caused by the introduction of contaminants into water intended to be used for drinking water, including an examination of the effectiveness of various drinking water technologies in removing, inactivating, or neutralizing biological, chemical, and radiological contaminants.
(6) Biomedical research into the short-term and long-term impact on public health of various chemical, biological, and radiological contaminants that may be introduced into public water systems through terrorist or other intentional acts

(b) Funding—

For the authorization of appropriations to carry out this section, see section 1435(e).

SECTION 1435. SUPPLY DISRUPTION PREVENTION, DETECTION, AND RESPONSE

(a) Distribution of Supply or Safety—

The Administrator, in coordination with the appropriate departments and agencies of the federal government, shall review or enter into contracts or co-operative agreements to provide for a review of methods and means by which terrorists or other individuals or groups could disrupt the supply of safe drinking water or take other actions against water collection, pretreatment, treatment, storage, and distribution facilities which could render such water significantly less safe for human consumption, including each of the following:

(1) Methods and means by which pipes and other constructed conveyances utilized in public water systems could be destroyed or otherwise prevented from providing adequate supplies of drinking water meeting applicable public health standards.
(2) Methods and means by which collection, pretreatment, treatment, storage, and distribution facilities utilized or used in connection with public water systems and collection and pretreatment storage facilities used in connection with public water systems, could be destroyed or otherwise prevented from providing adequate supplies of drinking water meeting applicable public health standards.
(3) Methods and means by which pipes, constructed conveyances, collection, pre-treatment, treatment, storage, and distribution systems that are utilized in connection with public water systems could be altered or affected so as to be subject to cross-contamination of drinking water supplies.
(4) Methods and means by which pipes, constructed conveyances, collection, pre-treatment, treatment, storage, and distribution systems that are utilized in connection with public water systems could be reasonably protected from terrorist attacks or other acts intended to disrupt the supply or affect the safety of drinking water.
(5) Methods and means by which information systems, including process controls and supervisory control and data acquisition and cyber systems at community water systems could be disrupted by terrorists or other groups.

(b) Alternative Sources—

The review under this section shall also include a review of the methods and means by which alternative supplies of drinking water could be provided in the event of the destruction, impairment or contamination of public water systems.

(c) Requirements and Considerations—

In carrying out this section and section 1434—

(1) the Administrator shall ensure that reviews carried out under this section reflect the needs of community water systems of various sizes and various geographic areas of the United States; and

(2) the Administrator may consider the vulnerability of, or potential for forced interruption of service for, a region or service area, including community water systems that provide service to the National Capital area.

(d) Information Sharing—

As soon as practicable after reviews carried out under this section or section 1434 has been evaluated, the Administrator shall disseminate, as appropriate as determined by the Administrator, to community water systems information on the results of the project through the Information Sharing and Analysis Center, or other appropriate means.

(e) Funding—

There are authorized to be appropriated to carry out this section and section 1434 not more than $15,000,000 for the fiscal year 2002 and such sums as may be necessary for the fiscal years 2003 through 2005.

Index

About the Author

Frank R. Spellman is assistant professor of environmental health at Old Dominion University in Norfolk, Virginia. He has extensive experience (more than thirty-five years) in environmental science and engineering, in both the military and the civilian communities. A professional member of the American Society of Safety Engineers, the Water Environment Federation, and the Institute of Hazardous Materials Managers, Frank Spellman is a board certified safety professional and a board certified hazardous materials manager. He has authored or coauthored 48 texts on safety, occupational health, water/wastewater operations, environmental science, and concentrated animal feeding operations.